SpringerBriefs in Physics

SpringerBriefs in Physics are a series of slim high-quality publications encompassing the entire spectrum of physics. Manuscripts for SpringerBriefs in Physics will be evaluated by Springer and by members of the Editorial Board. Proposals and other communication should be sent to your Publishing Editors at Springer.

Featuring compact volumes of 50 to 125 pages (approximately 20,000–45,000 words), Briefs are shorter than a conventional book but longer than a journal article. Thus, Briefs serve as timely, concise tools for students, researchers, and professionals.

Typical texts for publication might include:

- A snapshot review of the current state of a hot or emerging field
- A concise introduction to core concepts that students must understand in order to make independent contributions
- An extended research report giving more details and discussion than is possible in a conventional journal article
- A manual describing underlying principles and best practices for an experimental technique
- An essay exploring new ideas within physics, related philosophical issues, or broader topics such as science and society

Briefs allow authors to present their ideas and readers to absorb them with minimal time investment. Briefs will be published as part of Springer's eBook collection, with millions of users worldwide. In addition, they will be available, just like other books, for individual print and electronic purchase. Briefs are characterized by fast, global electronic dissemination, straightforward publishing agreements, easy-to-use manuscript preparation and formatting guidelines, and expedited production schedules. We aim for publication 8–12 weeks after acceptance.

Ralf Blossey

The Poisson-Boltzmann Equation

An Introduction

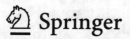 Springer

Ralf Blossey
CNRS
University of Lille
Villeneuve d'Ascq Cedex, France

ISSN 2191-5423   ISSN 2191-5431 (electronic)
SpringerBriefs in Physics
ISBN 978-3-031-24781-1   ISBN 978-3-031-24782-8 (eBook)
https://doi.org/10.1007/978-3-031-24782-8

This Springer imprint is published by the registered company Springer Nature Switzerland AG
The registered company address is: Gewerbestrasse 11, 6330 Cham, Switzerland

To students, colleagues, friends and family

Preface

This short book is intended to provide a concise introduction into *Poisson-Boltzmann theory* (often short: PB theory). Theory seems a big word here, and the first question asked may be whether Poisson-Boltzmann theory actually *is* a theory, or rather a modeling approach. The readers will find my answer upon reading and working through the book.

What is *Poisson-Boltzmann theory* about? The readers will see that what this book contains (to lowest order) is essentially a set of tools that allow to solve the Maxwell equation

$$\nabla \cdot \mathbf{D} = \varrho$$

whereby $\mathbf{D}$ is the dielectric displacement field, and ϱ the density of charges. We will mostly consider simple geometries, typically of planar type, like a single wall or a slit or channel containing electric charges with density ϱ that will be both fixed, typically on the system boundaries, and 'mobile'—dissolved charges in the bulk liquid—a prototypical *electrolyte*. In the latter case, the charge density ϱ generally turns into a nonlinear function of the electrostatic potential, and the ensuing equation becomes an involved differential equation of the electrostatic potential.

This very basic system setup is, at the same time, of an incredible generality in soft matter, physico-chemical and biophysical systems. Electrolyte systems bounded by charged surfaces are in a sense 'elementary building blocks' in these contexts. On the other hand, despite this generality, they are also of an enormous richness and diversity, necessitating extensions of the basic Poisson-Boltzmann theory which has not seen an end, and most likely won't.

The material in this book is arranged in three chapters. Chapter 1 introduces the Poisson-Boltzmann equation straightaway for the case of (1:1) salt, i.e., an electrolyte containing dissolved monovalent ions, e.g., the case of sodium chloride, NaCl, dissolving in Na^+ and Cl^-. Here we discuss the relevant physical lengths scales and the behavior of the solutions to the PB equation in planar, cylindrical and spherical geometries. We also show how Poisson-Boltzmann equations can be obtained for more complex fluids with involved equations of state, and we will show that spatial density variations, as they occur, e.g., in ionic liquids, give rise to Poisson-Boltzmann equations of higher order.

Chapter 2 goes one step further. Here, we now derive the Poisson-Boltzmann equation from a systematic statistical physics approach which allows us to show that it is the saddle point of the field-theoretic partition function. This approach will enable us to go beyond mean field and consider the one-loop correction to its solution. As an application, we compute the surface tension of an air/electrolyte interface, as a classic problem in the field of soft matter electrostatics which was first discussed in the 1930s. We then introduce a variational method that allows to derive a fluctuation-corrected Poisson-Boltzmann equation and provide a first illustrative application of the method for the interaction of a charged polymer with a like-charged membrane.

Chapter 3 addresses the problem of going beyond the structureless solvent described solely by its macroscopic dielectric constant in PB theory. To make the solvent properties 'explicit', we first introduce a phenomenological approach by introducing a wave vector-dependent dielectric function or permittivity. Subsequently, we discuss a microscopic model of a Poisson-Boltzmann equation with explicit solvent modeled by point dipoles, before turning the point dipoles into finite-size molecules. This approach allows us to determine, e.g., a model effective dielectric function, first in a mean-field approach. We then apply the variational approach from Chap. 2 to a Hamiltonian adapted to the slit geometry already discussed in standard Poisson-Boltzmann theory in Chap. 1. Ultimately, we provide solutions to the variational equations for this Hamiltonian in special cases and discuss the resulting physics. And, finally, we conclude on the contents of this book.

The author of this book of course hopes that it will be useful to its readers who wish to start with Poisson-Boltzmann theory but also like to go beyond the most basic levels already. An understanding of thermodynamics and statistical physics at an advanced level is required. Working through it should be seen as a mountaineering effort: the challenge grows with each chapter, sometimes some more easygoing paths emerge, but then it might become steep again. Reaching the top is a reward well worth taking the path. For those who are willing to even look on for the next challenge, I have provided several suggestions for further reading at the end of each chapter.

Lille, France   Ralf Blossey
November 2022

Acknowledgements

It is a pleasure to thank my collaborators and colleagues for shaping my view of the subject. I profited from many discussions over the years with Hélène Berthoumieux, Markus Bier, Sahin Buyukdagli, Fabrizio Cleri, Guillaume Copie, Marc Delerue, Ralf Everarers, Andreas Hildebrandt, Marc Lensink, Anthony Maggs, Arghya Majee, Roland Netz, Henri Orland, Fabien Paillusson, Rudi Podgornik, Emmanuel Trizac and many others. Andreas Hildebrandt is to be particularly credited with the work on phenomenological nonlocal electrostatics performed during his Ph.D. at Saarland University, as is Sahin Buyukdagli for results on the solvent-explicit dipolar nonlocal electrostatics discussed in this book, obtained during an extraordinarily productive postdoctoral stay in Lille.

Contents

Acronyms

A	Hamaker constant
β	$= 1/k_B T$, inverse thermal energy
ℓ_B	Bjerrum length
C	Differential capacitance
$\mathbf{D(r)}$	Dielectric displacement field
$\mathcal{D}$	Path integral integration measure
ε_0	Dielectric constant of vacuum
ε_r	Relative dielectric constant
$\varepsilon_\parallel$	Parallel dielectric permittivity
$\varepsilon_\perp$	Perpendicular dielectric permittivity
$\varepsilon(k)$	Wave-vector dependent dielectric function
$\mathbf{E(r)}$	Electric field
F	Free energy
F_{fl}	Fluid free energy
F_{el}	Electrostatic free energy
$f(c)$	Free energy density
G, v	Green function
$\Delta\gamma, \Gamma$	Surface tension
κ, κ_D	Debye screening parameter
H	Hamiltonian
λ_D	Debye length
$\lambda_\pm, \Lambda_{pm}$	Fugacity
$\mathcal{L}$	Langevin function
ℓ_{GC}	Gouy-Chapman length
μ	Chemical potential
n	Number density
ϕ, ψ	Electrostatic potential
Ω	Grand potential
Π	Osmotic pressure
q, Q	Electric charge
ϱ	Charge density

σ, Σ	Surface charge density
τ	Line charge density
v_0	Covariance, see Green function
$S(k)$	Structure factor
Ξ, Γ	Coupling parameter
Z	Partition function

Chapter 1
The Poisson-Boltzmann Equation

The first chapter of the book discusses the formulation and the solutions of the Poisson-Boltzmann equation for simple salts (like sodium chloride, NaCl) in simple geometries like planar, cylindrical and spherical systems. Special emphasis is laid on the role of a number of *physical length scales* characterizing the solutions. Subsequently, we will see that the Poisson-Boltzmann equation can be generalized for different fluid systems, and that its can also turn into a higher-order differential equation if density variations in a binary fluid are allowed.

1.1 Length Scales

Poisson-Boltzmann theory is very much a theory of *length scales*; we will try to explain and illustrate this in the first section of the first chapter, making in part use of Ref. [1].

Figure 1.1 illustrates the basic situation we will address in the following for the solution of the *Maxwell equation*

$$\nabla \cdot \mathbf{D} = \varrho. \tag{1.1}$$

An electrolyte solution is composed of a *simple salt* with the charge ratio (1:1)—consider sodium chloride NaCl, which dissolves in a polar solvent into to Na^+ and Cl^-. The polar solvent is characterized by the dielectric constant ε. Let us make a point on physical units right away. In the above formula, we make use of the SI system of units. The dielectric constant ε is thus given by the expression

$$\varepsilon = \varepsilon_r \varepsilon_0 \tag{1.2}$$

in which ε_r is the *relative dielectric constant* with respect to the *dielectric constant of vacuum*, $\varepsilon_0 \approx 8.85 \times 10^{-12}$ As/Vm. In the other often used system of units in electrostatics, Gaussian units, one sets

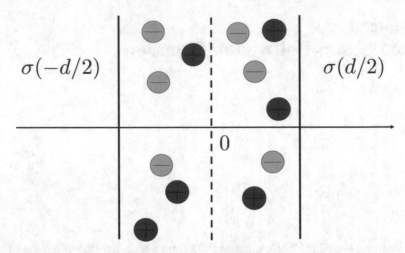

Fig. 1.1 Slit of width d with the electrolyte placed between two walls, here placed at $z = -d/2$ and $z = d/2$. The solvated ions are shown as red and green spheres carrying positive and negative charges. In the basic Poisson-Boltzmann model, the ions have no extension: they are considered as point charges. The walls carry surface charge distributions $\sigma(-d/2)$ and $\sigma(d/2)$, respectively

$$4\pi\varepsilon_0 = 1. \tag{1.3}$$

Although the schematic figure indicates a finite size of the ions—and does not represent the solvent molecules at all—our theory will at first ignore the physical size of both the ions and the solvent molecules. Considering the system at a given temperature T we can define a first characteristic length scale, the *Bjerrum length* ℓ_B. It corresponds to the condition of the equality of the electrostatic (i.e., Coulomb) energy between two charges $q = Ze$ with the thermal energy $k_B T$, wherein e is the elementary charge and Z the charge valency. Setting these contributions equal means putting

$$k_B T = \frac{q^2}{4\pi\varepsilon\ell_B} \tag{1.4}$$

with k_B as the Boltzmann constant. This expression gives for the Bjerrum length the result

$$\ell_B \equiv \frac{\beta q^2}{4\pi\varepsilon}, \tag{1.5}$$

with $\beta \equiv 1/(k_B T)$. We therefore have the behaviours $\ell_B \sim 1/T$ and $\ell_B \sim 1/\varepsilon$: for both high temperatures and strongly polar liquids (like water with $\varepsilon_r = 80$), the Bjerrum length is a small parameter. For a monovalent salt in water at room temperature, $\ell_B \approx 0.7$ nm [1].

Making use of the Bjerrum length we can define a second characteristic length, the *Debye length* or *Debye screening length* λ_D. The notion of screening refers to the fact that dissolved (mobile) charges in the electrolyte dampen the electric field applied at the surface. Considering not just a pair of interacting charges, but imagining those pairs as being imbedded in a volume characterized by an ionic density of positive and negative charges $n_+ + n_- = n_0$, we can define the screening length λ_D via the relation

$$n_0 \sim \frac{1}{\ell_B} \cdot \frac{1}{\lambda_D^2}. \tag{1.6}$$

Taking the proportionality factor as 4π (for convenience, as it cancels the 4π-factor in the denominator of the Bjerrum length), we obtain

$$\lambda_D \equiv \frac{1}{\sqrt{4\pi \ell_B n_0}}. \tag{1.7}$$

The *Debye length* depends inversely on salt concentration: the higher the concentration, the shorter the screening length. Often the inverse of the Debye length, κ or κ_D is used; the *differential capacitance C* defined by

$$C \equiv \frac{d\sigma}{d\phi} \tag{1.8}$$

where σ is the *surface charge density* is related to this quantity via $C \sim \kappa$; we will discuss this relationship in Chap. 3.

Finally, from the interaction of a charge with a uniformly charged surface with surface charge density σ we can define the *Gouy-Chapman length*

$$\ell_{GC} \equiv \frac{1}{2\pi \ell_B \sigma}. \tag{1.9}$$

We can thus also write

$$\sigma = \frac{1}{2\pi \ell_B \ell_{GC}}. \tag{1.10}$$

The three lengths, the Bjerrum length ℓ_B, the Debye length λ_D and the Gouy-Chapman length ℓ_{GC} are the key length parameters for the electrostatics of a confined electrolyte like the one sketched in Fig. 1.1.

We now move on to write down the differential equation governing the electrostatic potential in the electrolyte.

1.2 The Poisson-Boltzmann Equation for (1:1) Salt

In order to provide this derivation, we return to the Maxwell equation

$$\nabla \cdot \mathbf{D}(\mathbf{r}) = \varrho(\mathbf{r}) \tag{1.11}$$

where we have $\mathbf{D}(\mathbf{r}) = \varepsilon \mathbf{E}(\mathbf{r})$ on the left-hand side of the equation. For its right-hand side, we need to specify the charge density in the bulk. Assuming a (1:1) charge ratio—monovalent salt—we have charges $q \equiv \pm e$ with number densites $n_\pm(\mathbf{r})$. Thus we have, as before

$$n(\mathbf{r}) = n_+(\mathbf{r}) + n_-(\mathbf{r}) \tag{1.12}$$

and with $\varrho(\mathbf{r}) = qn(\mathbf{r})$ one finds

$$\varrho(\mathbf{r}) = e[n_+(\mathbf{r}) - n_-(\mathbf{r})]. \tag{1.13}$$

The number densities of positive and negative charges are assumed to be controlled by a *Boltzmann factor* involving the electrostatic potential ψ, fulfilling $\mathbf{E}(\mathbf{r}) = -\nabla \psi(\mathbf{r})$ in the form

$$n_\pm(\mathbf{r}) = n_\pm e^{\mp \beta e \psi(\mathbf{r})}, \tag{1.14}$$

with $n_+ = n_- = n/2$. Introducing

$$\phi(\mathbf{r}) \equiv \beta e \psi(\mathbf{r}) \tag{1.15}$$

we obtain

$$\Delta \phi(\mathbf{r}) = \kappa^2 \sinh \phi(\mathbf{r}) \tag{1.16}$$

with the *Debye screening parameter* $\kappa = 1/\lambda_D$.

This is the *Poisson-Boltzmann equation for (1:1) salt*. This in general nonlinear partial differential equation will be the key topic of our discussion in this first chapter. We will only consider situations in which the symmetry of the system will allow us to reduce the partial differential equation to an ordinary differential equation, thereby considerably simplifying calculations.

1.3   Solution of the Poisson-Boltzmann Equation for the Single-Plate Geometry

In order to solve the Poisson-Boltzmann Eq. (1.16) we need to specify the geometry of the system and, correspondingly, the boundary conditions. We begin with a half-space system which is bounded by a *single wall* or plate carrying a surface charge density Σ. The equation then turns one-dimensional because of translational invariance in the two-dimensional (x, y)-plane. Thus we only need to consider the ordinary differential equation

$$\frac{d^2 \phi(z)}{dz^2} = \kappa^2 \sinh \phi(z) \tag{1.17}$$

that can be readily integrated by *quadrature*, i.e. by multiplication of the equation by $d\phi/dz$ and its formal integration. This yields the expression

$$\frac{1}{2}\left(\frac{d\phi(z)}{dz}\right)^2 = \kappa^2 \left(\cosh(\phi(z)) - 1\right), \tag{1.18}$$

where the last term in the equation stems from the (natural) boundary condition that the electrostatic potential vanishes at infinity,

$$\lim_{z\to\infty}\frac{d\phi}{dz} = \phi(\infty) = 0, \tag{1.19}$$

which fixes the value of the integration constant. Since $\cosh(\phi) = 1 + 2\sinh^2(\phi/2)$ one can obtain

$$\frac{d\phi(z)}{dz} = -2\kappa\sinh(\phi(z)/2), \tag{1.20}$$

with ϕ positive and its derivative being negative. A separation of variables leads to the expression

$$\frac{d\phi}{\sinh(\phi/2)} = -2\kappa\,dz \tag{1.21}$$

that can now be integrated on the intervals $[\phi_s, \phi]$ and $[0, z]$. The integrand of the inverse hyperbolic sine can be written as

$$\frac{1}{\sinh(\phi/2)} = \frac{1}{2}\frac{1}{\sinh(\phi/4)\cosh(\phi/4)}, \tag{1.22}$$

making use of an identity for hyperbolic functions. This expression can be further rewritten as, using the representation of unity for hyperbolic sines and cosines,

$$\frac{1}{2}\frac{\cosh^2(\phi/4) - \sinh^2(\phi/4)}{\sinh(\phi/4)\cosh(\phi/4)} = \frac{1}{2}\left(\frac{\cosh(\phi/4)}{\sinh(\phi/4)} - \frac{\sinh(\phi/4)}{\cosh(\phi/4)}\right) \tag{1.23}$$

$$= 2\frac{d}{d\phi}\left[\ln\left(\cosh(\phi/4)\right) - \ln\left(\sinh(\phi/4)\right)\right].$$

One thus obtains

$$\ln\left(\frac{\tanh(\phi/4)}{\tanh(\phi_s/4)}\right) = -\kappa z. \tag{1.24}$$

Solving this equation for ϕ, one obtains with the help of the identity

$$\operatorname{arctanh}(z) = \frac{1}{2}\ln\left(\frac{1+z}{1-z}\right) \tag{1.25}$$

the final expression

$$\phi(z) = 2 \ln \left(\frac{1 + \eta e^{-\kappa z}}{1 - \eta e^{-\kappa z}} \right) \tag{1.26}$$

where

$$\eta \equiv \tanh(\phi_s/4). \tag{1.27}$$

It now remains to determine the value of ϕ_s. From the derivative of the electrostatic potential at the surface $z = 0$ we have

$$\frac{d\phi(z)}{dz} \bigg|_{z=0} = -\frac{\Sigma}{\varepsilon} = -4\pi \ell_B \sigma \, \mathrm{sgn}(\Sigma) = -2 \frac{\mathrm{sgn}(\Sigma)}{\ell_{GC}}. \tag{1.28}$$

Employing the bulk equation for the derivative of ϕ we have the *Grahame equation*

$$\sinh(\phi_s/2) = \mathrm{sgn}(\Sigma) \frac{\lambda_D}{\ell_{GC}}, \tag{1.29}$$

from which the value of ϕ_s or the parameter η can be computed as

$$\phi_s \equiv \phi(z=0) = 2 \, \mathrm{sgn}(\Sigma) \ln \left(\frac{\lambda_D}{\ell_{GC}} + \left[1 + \frac{\lambda_D}{\ell_{GC}}^2 \right]^{1/2} \right) \tag{1.30}$$

and

$$\eta = \mathrm{sgn}(\Sigma) \frac{\lambda_D}{\ell_{GC}} \left(\left[1 + \frac{\lambda_D}{\ell_{GC}}^2 \right]^{1/2} - 1 \right), \tag{1.31}$$

where we have made ample use of the lengths introduced earlier in the first section.

In the case when the surface potential ϕ_s is small, $|\phi_s| \ll 1$, the Poisson-Boltzmann equation can be linearized, which is called the *Debye-Hückel limit*. The equation then reads as

$$\frac{d^2\phi(z)}{dz^2} = \kappa^2 \phi(z) \tag{1.32}$$

which is an exemplary case of the *Helmholtz equation* in which the Laplacian of a function is proportional to the function itself. In the present one-dimensional case it has the very simple solution of a purely exponential function

$$\phi(z) = \phi_s e^{-\kappa z}. \tag{1.33}$$

Following the Grahame equation, this relation requires

$$\phi_s \approx 2\mathrm{sgn}(\Sigma) \frac{\lambda_D}{\ell_{GC}} \ll 1, \tag{1.34}$$

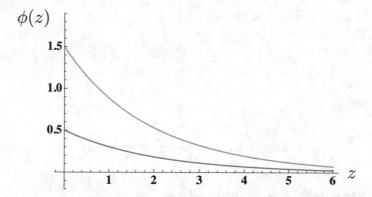

Fig. 1.2 The solution of the Poisson-Boltzmann equation for the single plate. Brown: the electrostatic potential of the nonlinear Poisson-Boltzmann equation, given by Eq. (1.26); blue: the Debye-Hückel solution, Eq. (1.33). Note that for a boundary value of the potential $\phi_s < 1$, both curves will overlay

hence the condition

$$\lambda_D \ll \ell_{GC}. \tag{1.35}$$

The meaning of this relation is that the Debye-Hückel behaviour requires that the charged-surface effects dominate over the screening of the ions in the bulk electrolyte. Figure 1.2 plots the electrostatic potential for the full nonlinear case and the Debye-Hückel linearized version of the Poisson-Boltzmann equation.

1.4 Slit Geometry

We are now sufficiently equipped to discuss the case of a *slit* or *pore* as shown in Fig. 1.1. We will assume positive and identical values of the surface potential ϕ_s, hence a symmetric configuration for the electrostatic potential. The solution to the Poisson-Boltzmann equation will then be monotonically decreasing from $z = -d/2$ towards $z = 0$ and then rising up towards $z = d/2$, passing through the minimum at the origin $z = 0$.

Our result from the quadrature of the nonlinear Poisson-Boltzmann equation will thus continue to hold provided we adopt the boundary condition accordingly. Supposing we solve the equation on the positive interval $[0, d/2]$ with the corresponding potential values $[\phi_m, \phi_s]$, we have

$$\frac{1}{2}\left(\frac{d\phi(z)}{dz}\right)^2 = \kappa^2 \left(\cosh(\phi(z)) - \cosh(\phi_m)\right). \tag{1.36}$$

The boundary conditions then read as

$$\left.\frac{d\phi}{dz}\right|_{z=\frac{d}{2}} = \frac{2}{\ell_{GC}}, \quad \left.\frac{d\phi(z)}{dz}\right|_{z=0} = 0, \tag{1.37}$$

and we find

$$\kappa z = \int_{\phi_m}^{\phi} d\phi' \frac{1}{\sqrt{2(\cosh\phi' - \cosh\phi_m)}}. \tag{1.38}$$

The boundary condition reads, accordingly, as

$$2\frac{\lambda_D^2}{\ell_{GC}^2} = \cosh\phi_s - \cosh\phi_m. \tag{1.39}$$

The expression (1.38) is an *elliptic integral* which cannot be given in more elementary functions, but rather in its own dedicated functions. Here we will therefore resort to an approximate treatment. Assuming large values of $\phi \gg 1$, we can simplify the hyperbolic cosine in the form

$$\cosh x \approx e^x/2$$

neglecting the decaying exponential. We then rewrite Eq. (1.38) as

$$\kappa z = \int_{\phi_m}^{\phi} d\phi' \left[e^{\phi'} - e^{\phi_m}\right]^{-1/2}. \tag{1.40}$$

The integration variable can be transformed via $x \equiv \phi - \phi_m$ leading to

$$\kappa z = e^{-\phi_m/2} \int_{0}^{\phi-\phi_m} dx \frac{1}{\sqrt{e^x - 1}}. \tag{1.41}$$

The integral is given by the trigonometric function

$$\arctan\left(\sqrt{e^{\phi-\phi_m} - 1}\right) \tag{1.42}$$

so that with further trigonometric relations one arrives at the expression

$$e^{-\frac{\phi-\phi_m}{2}} = \cos\left(\kappa z e^{\phi_m/2}\right) \tag{1.43}$$

that can now be solved to obtain the electrostatic potential ϕ,

$$\phi(z) = \ln\left[\frac{e^{\phi_m}}{\cos(\kappa z e^{\phi_m/2})}\right]. \tag{1.44}$$

We have thus obtained an approximate solution of the Poisson-Boltzmann equation, valid in the limit $\phi \gg 1$.

As before in the case of the single plate we can also solve the linearized Poisson-Boltzmann equation for the slit problem, leading to the corresponding Debye-Hückel result. It reads as

$$\phi(z) = \phi_m \cosh(\kappa z), \tag{1.45}$$

from which the two relations

$$\phi_s = \frac{2}{\kappa \ell_{GC}} \coth(\kappa d/2) \tag{1.46}$$

and

$$\phi_m = \frac{\phi_s}{\cosh(\kappa d/2)} \tag{1.47}$$

follow.

While it is nice to have determined the electrostatic potential analytically, it is even better to determine a measurable physical quantity from it. In the two-plate configuration the quantity of interest is the *osmotic pressure* Π acting on the electrolyte solution between the two plates. It follows from the first integral of the PB equation; see the detailed discussion in Refs. [2, 3]. In our case it is given by

$$\Pi = -\frac{k_B T}{8\pi \ell_B}\left(\frac{d\phi(z)}{dz}\right)^2 + 2k_B T n_0 \left(\cosh\phi(z) - 1\right) \tag{1.48}$$

which is a constant. It is thus advantageous to compute it at the midpoint of the two-plate system, where it reduces to

$$\Pi = 2k_B T n_0 \left(\cosh\phi_m - 1\right). \tag{1.49}$$

Since we have discussed either large or small values of ϕ_m before, we have two limiting cases. For large ϕ_m, we have

$$\Pi = k_B T n_0 e^{\phi_m}, \tag{1.50}$$

with the same approximation as before. For small ϕ_m we have

$$\Pi = k_B T n_0 \phi_m^2. \tag{1.51}$$

What remains to do now is to calculate the expressions of ϕ_m in the corresponding regions, making use of the lengths we have defined at the beginning of the chapter. In determining ϕ_m we will focus in the following on the dependency on three lengths: the plate distance d, the Debye screening length λ_D, which characterizes the bulk electrolyte, and the Gouy-Chapman length ℓ_{GC}, which characterizes the effect of the surface charges into the bulk electrolyte. We start with the limit of small d. To be more precise, we define

Regime 1: $d \ll \lambda_D, d \ll \ell_{GC}$.

In this regime, one starts with the relation

$$\left(\frac{\lambda_D^2}{\ell_{GC}^2} \right) \sim e^{\phi_s} - e^{\phi_m}, \tag{1.52}$$

obtained from Eq. (1.39). Note that here and in the following we will ignore precise numerical factors to work out the essential length dependencies (basic *scaling behaviours*) with respect to the system lengths of interest. From Eq. (1.43) for $\phi = \phi_s$ we obtain a second relation for e^{ϕ_s} which allows us to obtain a closed equation for ϕ_m. In fact, we have

$$e^{\phi_s} = \frac{e^{\phi_m}}{\cos^2 \left(\frac{d}{2\lambda_D} e^{\phi_m/2} \right)}. \tag{1.53}$$

Inserting this equation into the previous one, using a trigonometric identity and expanding the argument one finds

$$\left(\frac{\lambda_D^2}{\ell_{GC}^2} \right) \sim e^{\phi_m} \frac{d^2}{\lambda_D^2} e^{\phi_m}, \tag{1.54}$$

from which

$$e^{2\phi_m} \sim \frac{\lambda_D^4}{\ell_{GC}^2 d^2} \tag{1.55}$$

results. We finally have the scaling result

$$\Pi(d) \sim \frac{1}{\lambda_D^2} e^{\phi_m} = \frac{1}{\ell_{GC} d}. \tag{1.56}$$

This regime is called the *ideal gas regime*.

Regime 2: $d \ll \lambda_D; d \gg \ell_{GC}$.

In this second regime, we have to look at the zero of the cos-function in Eq. (1.53). There

$$\frac{d}{\lambda_D} e^{\phi_m/2} \approx \pi. \tag{1.57}$$

Thus we directly have

$$\Pi(d) \sim \frac{1}{\lambda_D^2} e^{\phi_m} = \frac{1}{d^2}. \tag{1.58}$$

This is the *Gouy-Chapman regime*.

The next regime follows for large plate separations d:

Regime 3: $d \gg \lambda_D$; $d \gg \ell_{GC}$.

Here we have $\phi_m \sim e^{-d/2\lambda_D}$. With the quadratic dependence of the osmotic pressure on ϕ_m we obtain the scaling

$$\Pi(d) \sim e^{-d/\lambda_D}. \tag{1.59}$$

This regime is called the *intermediate regime*.

Finally we have to consider the regime

Regime 4: $d \gg \lambda_D$; $d \ll \ell_{GC}$.

This is the regime of large plate separations which evidently corresponds to the *Debye-Hückel regime*. Here we have from Eqs. (1.46), (1.47) the expression

$$\phi_m \sim \frac{\lambda_D}{\ell_{GC}} \frac{1}{\sinh(d/2\lambda_D)} \tag{1.60}$$

and hence

$$P(d) \sim \frac{1}{\ell_{GC}^2} \frac{1}{\sinh^2(d/2\lambda_D)}. \tag{1.61}$$

All these regimes can be summarized in a simple graph, see Fig. 1.3. A considerably more detailed discussion of the regimes, distinguishing additional details in the scaling regimes only roughly sketched here, can e.g. be found in Ref. [2].

Fig. 1.3 Sketch of the scaling regimes of the solutions of the Poisson-Boltzmann equation for the symmetric slit as given in Fig. 1.1

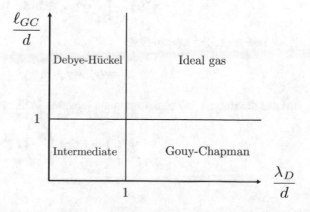

1.5 The Poisson-Boltzmann Equation in Cylindrical and Spherical Geometry

In this section we will solve the Poisson-Boltzmann equation for two other elementary geometries: a cylinder and a sphere. These cases generalize the case of the simple planar wall in an obvious way, and the relevance of these geometries is in fact easy to motivate. In the case of the cylinder one can just think of the electrostatic field around a straight macromolecule like DNA in a salt solution. The spherical case can e.g. be used to determine the field around a colloidal or other nanoparticle.

In order to be as general as reasonably possible in obtaining analytic solutions to the problem, in a first step we eliminate (or rather, hide) the dependence of the Poisson-Boltzmann equation on the Debye screening parameter κ. This is achieved by measuring length in terms of the Debye length via $r \equiv \kappa x$. The PB equation then reads as

$$\Delta\phi(r) = \sinh(\phi(r)) \tag{1.62}$$

and since we consider the cylindrical and spherical geometry, we can write the Laplacian Δ as

$$\left(\frac{1}{r}\right)^n \frac{d}{dr}\left(r^n \frac{d}{dr}\right)\phi(r) = \sinh(\phi(r)). \tag{1.63}$$

This equation now covers the planar, cylindrical and spherical cases for $n = 0, 1, 2$, respectively.

It turns out that while the planar Poisson-Boltzmann equation admits exact analytical solutions, this is only also true for the cylindrical case, which we will therefore treat in more detail. In the spherical case, we will again need to look at simplified, asymptotic solutions.

We start with the spherical case, $n = 2$, and we have from Eq. (1.63)

$$\phi''(r) + \frac{2}{r}\phi'(r) = \sinh(\phi(r)). \tag{1.64}$$

If we now make the substitution

$$\psi(r) = r\phi(r) \tag{1.65}$$

we can rewrite the Poisson-Boltzmann equation in terms of ψ in the form

$$\psi''(r) = r \sinh\left(\frac{\psi}{r}\right). \tag{1.66}$$

Expanding the sinh-function this expression is equivalent to

$$\psi''(r) = \psi(r) + \frac{1}{3!}\frac{\psi(r)^3}{r^2} + \cdots \tag{1.67}$$

This expansion obviously works for $|\psi| < 1$, for which asymptotically all higher order terms will go to zero for large r. Thus, not surprisingly, we are back to the Debye-Hückel case

$$\psi''(r) = \psi(r) \tag{1.68}$$

with the general solution

$$\psi(r) = Ae^r + Be^{-r}. \tag{1.69}$$

The exponentially growing solution will violate the natural boundary condition $\psi(r \to \infty) = 0$ so that we need to require $A = 0$ and the solution to the spherical Poisson-Boltzmann equation will asymptotically behave as

$$\phi(r) \approx B\frac{e^r}{r}. \tag{1.70}$$

Now let's see what happens in the cylindrical case, $n = 1$. Keeping, as before, only the lowest order term and multiplying the equation by r^2, we have

$$r^2\phi''(r) + r\phi'(r) - r^2\phi(r) = 0, \tag{1.71}$$

which is the differential equation fulfilled by the *modified Bessel function of order zero*. It has the general solution

$$\phi(r) = c_1 I_0(r) + c_2 K_0(r), \tag{1.72}$$

where, as in the previous spherical case, the two Bessel functions I_0 and K_0 are exponentially growing and falling. Thus we have

$$\phi(r) \approx c_2 K_0(r) \tag{1.73}$$

which for $r \to \infty$ yields the leading order term

$$\phi(r) \approx C\frac{e^{-r}}{\sqrt{r}}. \tag{1.74}$$

The difference in the asymptotic behaviour of the cylindrical and spherical solutions thus lies in the power-law behaviour modifying the exponential decay that we can summarize as

$$\phi(r) \sim \frac{1}{r^{n/2}} e^{-r} \tag{1.75}$$

for all n.

Having established the asymptotic behaviour of the cylindrical and spherical Poisson-Boltzmann equations, we will now attack the analytic solution to the cylindrical case, following Ref. [4]. The nonlinear ordinary differential equation presents us with two problems, the highly nonlinear sinh-function and the coordinate singularity at the origin. Therefore it is useful to first transform the nonlinearity into a less complicated expression. This is, e.g., provided by the choice

$$\phi(r) = \ln \eta(r). \tag{1.76}$$

The logarithm will help us to get rid of the exponential function of which the sinh-function is composed. The resulting transformed equation reads as

$$\eta''(r) = \frac{1}{\eta(r)} (\eta'(r))^2 - \frac{1}{r} \eta'(r) + \frac{\eta^2(r) - 1}{2}. \tag{1.77}$$

In the next step we treat the singularity at $r = 0$. For this we transform to a variable t which relates to r via

$$e^t = \frac{r^2}{8}. \tag{1.78}$$

(The value of 8 is a judicious choice, but this only the final result conveys.) Note that here and in what follows we use the prime symbol $'$ to refer to the differentiation with respect to the argument of the given function. After this coordinate transformation the resulting equation reads as

$$\eta''(t) = \frac{1}{\eta(t)} (\eta'(t))^2 + e^t(\eta^2(t) - 1). \tag{1.79}$$

We can further take $z = e^t$ and are then left with

$$\eta''(z) = \frac{1}{\eta(z)} (\eta'(z))^2 - \frac{1}{z} (\eta'(z) - \eta^2(z) + 1). \tag{1.80}$$

This equation is a special case of the general class of *Painlevé equations* (in fact, it is called a Painlevé equation of the third kind). Its solutions have been shown to have no branch points or essential singularities in the z-plane; in the language of complex analysis they are meromorphic functions. In particular, the resulting analyticity of the solution function allows us to obtain it, while not in closed form, but nevertheless in the form of a convergent infinite series. Taking the version of Eq. (1.79), we take as the initial point for our expansion the point t_0, and expand in $v \equiv t - t_0$. We have

$$\eta(v)\eta''(v) = (\eta'(v))^2 + ke^v \eta(v)(\eta^2(v) - 1) \tag{1.81}$$

with $k = e^{t_0}$. Its solution is thus given by

$$\eta(v) = \sum_{n=0}^{\infty} a_n v^n \tag{1.82}$$

where from the insertion of this ansatz the expansion coefficients fulfill the recursion relation

$$a_2 = \frac{1}{2a_0}[a_1^2 + ka_0(a_0^2 - 1)] \tag{1.83}$$

and

$$a_{n+2} = \frac{1}{(n+2)(n+1)a_0} \left\{ \sum_{i=0}^{n} \left[(i+1)(n-i+1)a_{i+1}a_{n-i+1} - \frac{ka_{n-i}}{i!} \right. \right. \tag{1.84}$$

$$\left. \left. + k \left(\sum_{j=0}^{n-i} a_j a_{i-j} \right) \left(\sum_{j=0}^{n-i} \frac{a_{n-i-j}}{j!} \right) \right] - \sum_{i=0}^{n-1} (i+2)(i+1)a_{i+2}a_{n-i} \right\}, \, n > 0.$$

The coefficients a_0 and a_1 remain free parameters which have to be determined from the boundary conditions; we do not continue this discussion here. For all practicalities, the analytic solution to the Poisson-Boltzmann equation for the cylindrical case remains an unwieldy mathematical result. However, it makes the point that the solution to the cylindrical Poisson-Boltzmann equation falls into a class of well-studied ordinary differential equations—which does not hold for the spherical case.

1.6 Generalized Poisson-Boltzmann Equations

In this section we start to go beyond the standard Poisson-Boltzmann equation that applies to an unstructured solvent. Here we show that, under fairly moderate assumptions on the character of the electrolyte, generalized Poisson-Boltzmann equations can be formulated that deviate substantially from this standard form. We present a formal derivation of such Poisson-Boltzmann equations in which the appearance of higher-order derivative terms is directly linked to the spatial variation in the concentration of the ions in the binary mixture. The equation governing the electrostatic potential appears, to lowest relevant order, as a fourth-order partial differential equation whose coefficients, however, in general depend on the non-electrostatic equation of state. Our reasoning in this chapter is on the mean-field level within a thermodynamic approach that goes beyond our ad-hoc treatment for the derivation of the Poisson-Boltzmann equation for a (1:1)-salt in Sect. 1.2.

We take as the starting point of our discussion a *free energy* expression for a charged, isothermal binary mixture that we assume to be composed of two contributions, following the original work of Ref. [5],

$$F = F_{fl} + F_{el}. \tag{1.85}$$

The expression characterizing the fluid, F_{fl}, is written as

$$F_{fl}[c_1, c_2] = \int d^3\mathbf{r} \left(f(c_1, c_2) - \mu_1 c_1 - \mu_2 c_2 \right) \tag{1.86}$$

with the *free energy density* $f(c_1, c_2)$, a function of the densities of the two components of the mixture, c_1 and c_2, and the chemical potentials μ_1 and μ_2 fulfilling

$$\mu_{1,2} = \frac{\partial f(c_1, c_2)}{\partial c_{1,2}}. \tag{1.87}$$

The *Legendre transform* of the free energy density is the thermodynamic pressure, or the *equation of state*:

$$f(c_1, c_2) - \frac{\partial f(c_1, c_2)}{\partial c_1} c_1 - \frac{\partial f(c_1, c_2)}{\partial c_2} c_2 = -p(c_1, c_2), \tag{1.88}$$

so that

$$F_{fl} = -\int d^3\mathbf{r}\, p(\mu_1, \mu_2). \tag{1.89}$$

Together with the electrostatic contribution in Eq. (1.85), the full free energy F is given by the expression

$$F[c_1, c_2, \mathbf{D}] = \int d^3\mathbf{r} \left(f(c_1, c_2) - \mu_1 c_1 - \mu_2 c_2 \right) \tag{1.90}$$
$$+ \int d^3\mathbf{r} \left(\frac{\mathbf{D}^2}{2\varepsilon} - \phi(\nabla \cdot \mathbf{D}) - e(z_1 c_1 - z_2 c_2) \right),$$

that we can rearrange into

$$F[c_1, c_2, \mathbf{D}] = \int d^3\mathbf{r} \left(f(c_1, c_2) - (\mu_1 - e z_1 \phi) c_1 - (\mu_2 - e z_2 \phi) c_2 \right) \tag{1.91}$$
$$+ \int d^3\mathbf{r} \left(\frac{\mathbf{D}^2}{2\varepsilon} - \phi(\nabla \cdot \mathbf{D}) \right).$$

Making use of the *Legendre transform*, we can rewrite F in terms of the electrostatic potential by minimizing the expression with respect to $\mathbf{D}$ in the form

$$F[\phi] = -\int d^3\mathbf{r} \left(\frac{\varepsilon}{2} (\nabla\phi)^2 + p(\mu_1 - e z_1 \phi, \mu_2 - e z_2 \phi) \right). \tag{1.92}$$

One can now, for any equation of state $p(\mu_1, \mu_2)$ or free energy expression $f(c_1, c_2)$, derive the corresponding Poisson-Boltzmann equation, with a proper substitution of the chemical potentials via

$$\mu_i \rightarrow \mu_i \mp ez_i\phi, \tag{1.93}$$

in the form

$$\varepsilon\Delta\phi = \frac{\partial}{\partial\phi}p(\mu_1 - ez_1\phi, \mu_2 + ez_2\phi). \tag{1.94}$$

The right-hand side of this equation can be rewritten via the *Gibbs-Duhem equation*

$$c_{1,2} = \frac{\partial p}{\partial\mu_{1,2}} \tag{1.95}$$

as

$$\frac{\partial}{\partial\phi}p(\mu_1 - ez_1\phi, \mu_2 + ez_2\phi) = -ez_1\frac{\partial p}{\partial\mu_1} + ez_2\frac{\partial p}{\partial\mu_2} = -e(z_1c_1 - z_2c_2) = \tilde{q} \tag{1.96}$$

where $\tilde{q}$ is the local charge density.

Equation (1.92), together with Eq. (1.96), defines a generalized expression for the Poisson-Boltzmann equation in which the choice of the equation of state determines its precise shape: obviously, many different choices are possible. Our expression for the Poisson-Boltzmann equation for (1:1) salt corresponds to the case in which

$$p(c_1, c_2) = k_BT(c_1 + c_2) \tag{1.97}$$

is the *van't Hoff equation of state*. The standard Poisson-Boltzmann equation for a (1:1) (or other types of simple salts) thus corresponds to the case of an electrolyte described as an ideal gas of ions in a dielectric medium with dielectric constant $\varepsilon \approx 80$ in the case of water. Different equations of state and the corresponding Poisson-Boltzmann equations have been studied in Ref. [5], e.g. for an *asymmetric lattice gas* and for an asymmetric *binary hard-sphere mixture* described by the *Carnahan-Starling equation of state*, a model fluid equation of state frequently used in applications.

1.7 A Higher-Order Poisson-Boltzmann Equation

Expression (1.92) maintains the general form of a 'standard' Poisson-Boltzmann equation which consists of the Poisson equation but with a more complex, potential-dependent charge density. We now demonstrate that also the Poisson form of the equation can undergo modifications and that one then can end up with a higher-order differential operator in the Poisson-Boltzmann equation.

The starting point of this derivation is a modified fluid term that is taken as the sum of a local free energy plus a squared density gradient contribution. Following Ref. [6] we have

$$F_{fl}[c_1, c_2] = \int d^3\mathbf{r} \left(f(c_1, c_2) - \mu_1 c_1 - \mu_2 c_2 + \frac{1}{2} \sum_{i,j=1}^{2} \kappa_{ij} \nabla c_i \nabla c_j \right). \quad (1.98)$$

In addition to the previous case we now first introduce a vector-valued variable for the density gradient given by

$$\mathbf{v}_i \equiv \nabla c_i \quad (1.99)$$

with an associated *Lagrange multiplier* $\mathbf{g}_i$. An integration by parts and regrouping of terms in Eq. (1.98) leads to the expression for the full free energy, analogously to the simpler expression in the previous section,

$$F[c_1, c_2, \mathbf{D}] = \int d^3\mathbf{r} \left(f(c_1, c_2) - \sum_{i,i=1}^{2} (\mu_i + ez_i\phi + \nabla \cdot \mathbf{g}_i)c_i \right) \quad (1.100)$$

$$+ \int d^3\mathbf{r} \left(\frac{\mathbf{D}^2}{2\varepsilon} + \nabla\phi \cdot \mathbf{D} + \frac{1}{2} \sum_{i,j=1}^{2} (\kappa_{ij}\mathbf{v}_i\mathbf{v}_j - \mathbf{g}_i \cdot \mathbf{v}_i) \right).$$

We again recognize that the *stationary point* of this functional with respect to the concentration leads to the Legendre transform of the fluid free energy, thus

$$F[\phi, \{\mathbf{g}_i\}] = - \int d^3\mathbf{r} \frac{1}{2} \sum_{ij=1}^{2} \kappa_{ij}^{-1} \mathbf{g}_i \mathbf{g}_j - \left(\frac{\varepsilon}{2}(\nabla\phi)^2 + P(\{\mu_i - \phi ez_i + \nabla \cdot \mathbf{g}_i\}) \right)$$

$$(1.101)$$

where P is the fluid pressure expressed as a function of the set of chemical potentials μ_i, as discussed before. It is important to stress that the Lagrange multipliers $\mathbf{g}_i$ enter as an argument to the pressure function that in general is a highly nonlinear function.

We will now apply this setup to one and two-plate problems. As we had already seen in the earlier sections, the variational equations found by taking derivatives of the free energy F, Eq. (1.101), have a simple first integral in one-dimensional geometries. This integral can be found using the standard construction of a *Hamiltonian* from a *Lagrangian*, taking the coordinate x as equivalent to the time in a particle system, while the conserved quantity is not an energy, but rather the total pressure in the fluid where external charges are absent. It can be derived as

$$p = -\varepsilon\frac{(\partial_x\phi)^2}{2} + P + \frac{g_i g_j}{2\kappa_{ij}} - (\partial_x g_i)c_i. \quad (1.102)$$

Far from any external sources, where the potential and the g_i become constant, this reduces simply to the neutral fluid pressure function P.

In the case of a *symmetric binary mixture* Eq. (1.101) reduces to

$$F[\phi, \{\partial_x g_i\}] = -\int dx \left[\frac{\varepsilon}{2}(\partial_x \phi)^2 + \frac{1}{2}\overline{g} \cdot \widehat{\kappa} \cdot \overline{g}^T + P(\{\mu_i - ez_i\phi + \partial_x g_i\}) \right]$$

(1.103)

where, explicitly, we have for the inverse matrix with $i = (+, -)$, $\kappa_{ij}^{-1} \equiv \widehat{\kappa}_{ij}$

$$\overline{g} \cdot \widehat{\kappa} \cdot \overline{g}^T = (g_+ \; g_-) \begin{pmatrix} \widehat{\kappa}_{++} & -\widehat{\kappa}_{+-} \\ -\widehat{\kappa}_{+-} & \widehat{\kappa}_{--} \end{pmatrix} \begin{pmatrix} g_+ \\ g_- \end{pmatrix}.$$

(1.104)

The standard expression for the pressure comes from a lattice gas model with cell size a and a given parameter v:

$$P = \frac{1}{a^3} \ln \left(1 + \frac{v}{2(1-v)} \left(e^{\beta m_+(x)} + e^{\beta m_-(x)} \right) \right)$$

(1.105)

with

$$m_\pm(x) = \pm ez\phi(x) + \partial_x g_\pm(x).$$

(1.106)

We vary the functional $F[\phi, \{\partial_x g_i\}]$ with respect to the three fields ϕ, g_+ and g_-. The variation with respect to ϕ yields

$$-\varepsilon \partial_x^2 \phi(x) + \frac{vze}{a^3} \left[\frac{e^{\beta m_+(x)} - e^{\beta m_-(x)}}{1 - v + v(e^{\beta m_+(x)} + e^{\beta m_-(x)})} \right] = 0,$$

(1.107)

while for the two components of $\overline{g}$ one has

$$\widehat{\kappa}_{++}g_+ - \widehat{\kappa}_{+-}g_- = \frac{v}{a^3}\partial_x \left[\frac{e^{\beta m_+(x)}}{1 - v + v(e^{\beta m_+(x)} + e^{\beta m_-(x)})} \right]$$

(1.108)

and

$$\widehat{\kappa}_{--}g_- - \widehat{\kappa}_{+-}g_+ = \frac{v}{a^3}\partial_x \left[\frac{e^{\beta m_-(x)}}{1 - v + v(e^{\beta m_+(x)} + e^{\beta m_-(x)})} \right].$$

(1.109)

It is instructive to first consider the linearized equations—we will see that they already contain the essential physics of this theory. We find

$$-\varepsilon \partial_x^2 \phi(x) + \frac{vze\beta}{a^3}(2ez\phi(x) + \partial_x g_+(x) - \partial_x g_-(x)) = 0,$$

(1.110)

while for the two components of $\overline{g}$ one has

$$\widehat{\kappa}_{++}g_+(x) - \widehat{\kappa}_{+-}g_-(x) = \frac{v\beta}{a^3}(ez\partial_x\phi(x) + \partial_x^2 g_+(x))$$

(1.111)

and

$$\widehat{\kappa}_{--}g_-(x) - \widehat{\kappa}_{+-}g_+(x) = \frac{\nu\beta}{a^3}\left(-ez\partial_x\phi(x) + \partial_x^2 g_-(x)\right). \tag{1.112}$$

Figure 1.4 shows the solutions to these equations for the case of a one-plate system with a constant surface potential, $\phi(0) = V$, for the parameters indicated in the Figure caption. In Fig. 1.4a, $\kappa_{++} = \kappa_{--}$, while the contrast between the two couplings is increased by a factor of ten in Fig. 1.4b. The presence of g_+ and g_- leads to oscillations in the profile of the electrostatic potential whose number, amplitude and extension into the bulk depend on the contrast between the coupling parameters.

The equation for the electrostatic potential ϕ looks similar to that of a mechanical spring in which the position x plays the role of time, with the damping term given by the functions $g_+(x)$ and $g_-(x)$, which in turn essentially obey coupled diffusion equations, with the electric field acting as a source or a sink term. It is therefore not surprising that, as a function of the coupling parameters $\widehat{\kappa}_{ij}$, the solution exhibits damped oscillations.

The variational equations can be reduced to only two fields if one assumes the symmetry condition $\widehat{\kappa}_{++} = \widehat{\kappa}_{--} \equiv \kappa$. We further set $\widehat{\kappa}_{+-} = 0$. Subtracting the equations in $g_{\pm}$ and introducing $g \equiv g_+ - g_-$ then yields a coupled system in $\phi(x)$, $g(x)$. Being linear equations, one can eliminate the field g in favor of ϕ and ends up with a forth-order equation in the electrostatic field

$$\tilde{\kappa}^2\partial_x^4\psi(x) - \partial_x^2\psi(x) + \psi(x) = 0, \tag{1.113}$$

where $\psi \equiv (2ez)\phi$, $\tilde{\kappa}^2 \equiv 2(ze)^2(\nu\beta)^2/(\kappa\varepsilon a^6)$ and the spatial coordinate x has been rescaled by a factor $\sqrt{\epsilon}$ with $\epsilon = \varepsilon a^3/(\nu\beta)$.

Equation (1.113) is a linear equation that has been discussed in Ref. [7] in the context of ionic liquids, i.e. systems consisting of solvated 'large' ions whose structure cannot be adequately described in a point-charge picture. We identify the coupling parameter $\tilde{\kappa}^2$ with the parameter δ_c^2 used in that work. Furthermore, the exact solution to the linear equation that is derived in the Supplementary Material of Ref. [7] can immediately be adapted to the present case by just making the simple replacement δ_c^2 with $\tilde{\kappa}^2$.

In the nonlinear case, one has to solve the coupled Eqs. (1.110)–(1.112). A simplification can again be made if we consider the anti-symmetric subspace

$$g_+(x) = -g_-(x) \equiv \overline{g}(x), \tag{1.114}$$

corresponding to antagonistic gradients in concentrations. In this case one has $m_+(x) = -m_-(x) \equiv m(x)$ and the resulting two equations for ϕ and g can be written as

$$-\varepsilon\partial_x^2\phi(x) + \frac{\nu ze}{a^3}\frac{\sinh(\beta m(x))}{1 + 2\nu\sinh^2(\beta m(x)/2)} = 0 \tag{1.115}$$

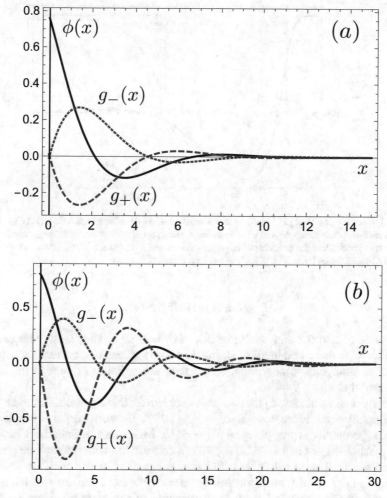

Fig. 1.4 Oscillatory profiles of the functions ϕ (black) g_+ (blue, large dashes) and g_- (red, small dashes), obtained from Eqs. (1.110), (1.111), (1.112) for two different sets of parameters. The parameters in (**a**) are $V = 0.8, \kappa_{++} = \kappa_{--} = 0.5, \kappa_{+-} = 0$. In (**b**): $V = 0.8, \kappa_{++} = 0.05, \kappa_{--} = 0.5, \kappa_{+-} = 0$. All other parameters are set to one. The other boundary conditions are $g_+(0) = g_-(0) = 0, \phi(L) = g_+(L) = g_-(L) = 0$. Figure taken from Ref. [6]; © American Physical Society

and

$$\overline{g}(x) = \frac{v}{2a^3\kappa} \partial_x \left(\frac{\sinh(\beta m(x))}{1 + 2v \sinh^2(\beta m(x)/2)} \right). \tag{1.116}$$

The full nonlinear equations can be solved numerically as a boundary value problem with the conditions $\phi(0) = V$, $\phi(L) = 0$, $\overline{g}(0) = \overline{g}(L) = 0$. Figure 1.5 shows the results for the charge density

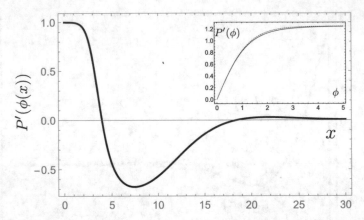

Fig. 1.5 The charge density $P'(\phi(x))$ from the solution of $\phi(x)$ obtained from Eqs. (1.115), (1.116) for the case of a constant potential $\phi(0) = 5$. Inset: the approximate expression used, compared to the exact expression of the pressure of the symmetric electrolyte (exact: black, approximation: red dashed). Figure taken from Ref. [6]; © American Physical Society

$$P'(\phi) \equiv (1/v)\tanh(v\phi(x)), \tag{1.117}$$

for a value of $v = 0.8$ and a value of $\phi(x = 0) = V = 5$. The results compare e.g. favorably with those obtained in [7], and therefore indicate that the explicit assumption of an antagonistic concentration gradient is implicitly present in the theory developed there.

Finally one can also study the two-plate problem for this equation. Figure 1.6 plots an exemplary osmotic pressure, given by Eq. (1.102), between the plates for identical constant potentials at the plates, $\phi(0) = \phi(L)$, based on the linearized two-field theory discussed just before. Also shown is a comparison with the osmotic pressure of the linearized standard Poisson-Boltzmann equation, which yields a parabolic potential $\phi(x) > 0$ with an exponentially decaying osmotic pressure for large plate distances. In the structured fluid, upon variation of the two-plate distance L, the potential ϕ first becomes flat at the center and crosses to negative values (as shown in an inset for $L = 10$), resulting in a pronounced minimum in $p(L)$ (not shown). Upon further increase in L the osmotic pressure progresses through a maximum and a second, very shallow minimum appears (see top inset). In this region, the electrostatic potential has a double-well structure next to the plates, as shown in the inset for $L = 15$. For still larger L, the double minima develop into a single minimum.

On can thus see that the inclusion of density gradients in a charged binary mixture leads in a natural way to the appearance of higher-order terms in the generalized Poisson-Boltzmann equation. In the model case of a symmetric binary mixture the presence of different coupling constants of the spatial gradient terms lead to pronounced oscillatory behaviour of the electrostatic profiles. Also, on the level of the generalized Poisson-Boltzmann equation, the oscillations near the wall therefore are the consequence of bulk behaviour and not of the surface coupling.

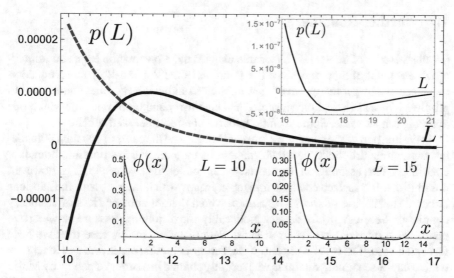

Fig. 1.6 Osmotic pressure in the linear two-field theory (black continuous line) and the linearized standard Poisson-Boltzmann equation (dashed blue line), both for constant potential $\phi(x=0) = 0.5$ at the plates. All parameters are set to one. In the range of plate distances L shown, the osmotic pressure shows oscillations between repulsive and attractive behaviors. The electrostatic potential, shown in two insets for the values of $L = 10$ and $L = 15$, evolve from having a single minimum at mid-plate to displaying two minima next to the plates. Figure taken from Ref. [6]; © American Physical Society

1.8 Summary

In this first chapter we have essentially discussed four topics. The first were the physical length scales involved in the solutions of the Poisson-Boltzmann equation for confined electrolyte systems. The second was the formulation of Poisson-Boltzmann equation itself and its analytic solutions for planar systems, a single wall and a double-wall or slit system. As our third topic, we discussed the solutions in the cylindrical and spherical cases. For the cylindrical case, we could show that the Poisson-Boltzmann equation can, after suitable transformations, be related to a class of ordinary differential equations studied by Painlevé. The spherical case, however, does not allow for such a treatment and its general solution remains so far unknown. The fourth and final part dealt with the derivation of generalized Poisson-Boltzmann equations, for which it was shown that this is possible in a straightforward manner for general equations of state which replace the hyperbolic sinh in the standard Poisson-Boltzmann equation. Further, it was shown that when considering spatial gradients in the fluid density, this amounts to the generation of higher-order derivatives in the Poisson-Boltzmann equation. We specifically studied the single- and double plate problem for a symmetric binary mixture which shows oscillatory behaviour of the electrostatic potential, a feature e.g. found in ionic liquids composed of large ions, for which the point-charge assumption cannot be justified anymore.

1.9   Further Reading

On the material of Sects. 1.1–1.7 of this chapter there meanwhile exists an unmanageable amount of literature, since the solutions of the Poisson-Boltzmann equation in planar or slit geometry have been applied in almost every imaginable way to problems from statistical physics, physical chemistry and biophysics for which this mapping can be made. Some specific remarks are nevertheless useful.

In the first part of the chapter we have only discussed the cases of constant potential or constant electric field (constant gradient of the potential) as possible boundary conditions. This assumes an inert surface that bounds the electrolyte. It is, however, possible that the surface contains ionizable groups which interact with the salt ions in bulk. This is case of *charge regulation* which defines another class of boundary conditions. See, e.g., [8] for details. Noticeably, the disjoining pressure between two plates has a different behaviour at short plate distances in this case than we have discussed here. The role of of charge regulation in membrane layers and stacks, as well as phase-separating mixtures was recently studied in a series of papers by Majee et al. [9–11]. The papers nicely illustrate the richness of the class of problems for which Poisson-Boltzmann theory is of importance.

On the problem of exact solutions of the Poisson-Boltzmann equation: the published papers on analytic solutions of the Poisson-Boltzmann equation are of a very different kind as compared to the application of this equation: their number is far smaller, and they are dispersed across journals of statistical physics, mathematical physics and physical chemistry. The most relevant work is certainly by Tracy and Widom [12] who exploited the link of the cylindrical Poisson-Boltzmann equation for (1:1) and (2:1) salts to *integrable systems of Painlevé/Toda type*. More recently, there have been a number of interesting papers by E. Trizac and collaborators on exact results for the screening of like charges by counterions [13], cylindrical polyions [14] and double layers with corrugated surface charge modulations [15]. Exact solutions remain a topic of interest, albeit a difficult one.

Finally, the last section in this chapter provides a link to the very active field of *ionic liquids*, in which theorists working on generalizations of Poisson-Boltzmann theories, density functional approaches, molecular dynamics simulations and, of course, experimental scientists work together. This field has literally exploded in the last years, not least because of its evident relevance for the developments of batteries and other applications of ionic liquids [16].

References

1. Herrero, C., Joly, L.: Poisson-Boltzmann Formulary. arXiv:2105.00720 (2021); updated (2022)
2. Andelman, D.: Electrostatic properties of membranes: the Poisson-Boltzmann theory. Handb. Biol. Phys. **1**, 603–642 (1995)
3. Adar, R.M., Andelman, D.: Osmotic pressure between arbitrarily charged planar surfaces: a revisited approach. Eur. Phys. J. E **41**, 11–18 (2018)

4. Benham, C.J.: The cylindrical Poisson-Boltzmann equation. I. Transformations and general solutions. J. Chem. Phys. **79**, 1969-1973 (1983)
5. Maggs, A.C., Podgornik, R.: General theory of asymmetric steric interactions in electrostatic double layers. Soft Matter **12**, 1219–1229 (2016)
6. Blossey, R., Maggs, A.C., Podgornik, R.: Structural interactions in ionic liquids linked to higher-order Poisson-Boltzmann equations. Phys. Rev. E **95**, 060602(R) (2017)
7. Bazant, M.Z., Storey, B.D., Kornyshev, A.A.: Double layer in ionic liquids: overscreening vs. crowding. Phys. Rev. Lett. **106**, 046102 (2011); Erratum Phys. Rev. Lett. **109**, 149903 (2012)
8. Markovich, T., Andelman, D., Podgornik, R.: Charge regulation: a generalized boundary condition? EPL **113**, 26004 (2016)
9. Majee, A., Bier, M., Podgornik, R.: Spontaneous symmetry breaking of charge-regulated systems. Soft Matter **14**, 985–991 (2018)
10. Majee, A., Bier, M., Blossey, R., Podgornik, R.: Charge regulation radically modifies electrostatics in membrane stacks. Phys. Rev. E **100**, 050601(R) (2019)
11. Majee, A., Bier, M., Blossey, R., Podgornik, R.: Charge symmetry broken complex coacervation. Phys. Rev. Res. **2**, 043417 (2020)
12. Tracy, C.A., Widom, H.: On exact solutions to the cylindrical Poisson-Boltzmann equation with applications to polyelectrolytes. Phys. A **244**, 402–413 (1997)
13. Téllez, G., Trizac, E.: Screening like-charges in one-dimensional coulomb systems: exact results. Phys. Rev. E **92**, 042134 (2015)
14. Téllez, G., Trizac, E.: Free energy of cylindrical polyions: analytical results. J. Chem. Phys. **151**, 124904 (2019)
15. Šamaj, L., Trizac, E.: Electric double layers with surface charge modulations: novel exact Poisson-Boltzmann solutions. Phys. Rev. E **100**, 042611 (2019)
16. Kornyshev, A.A.: Double-layer in ionic liquids: paradigm change? J. Phys. Chem B **111**, 5545–5557 (2007)

Chapter 2
Poisson-Boltzmann Theory and Statistical Physics

In the first chapter we went straight away to the Poisson-Boltzmann equation, formulating it by starting from the Maxwell equation. Such an approach is difficult to generalize and thus to improve upon (although we have made progress employing thermodynamics). The Poisson-Boltzmann equation is what is called a *mean-field equation*, which means it ignores thermal fluctuations that might be important in specific physical situations. In order to be able to account for fluctuations, we therefore need to extend our derivation.

In this second chapter we will derive the Poisson-Boltzmann equation from a statistical physics approach by first formulating the *partition function* of a system of ions interacting via the Coulomb potential—a '*Coulomb fluid*'. In this section we will further address a specific problem in detail, which is the computation of the *surface tension* γ of an electrolyte/air interface. This is a classic topic in the field, first addressed by Onsager and Samaras [1]. This section closely follows the work by Markovich, Andelman and Podgornik, published by the authors in two papers [2, 3]. The key point here is that beyond a slightly more involved mean-field solution to the PB-equation, we need to compute a fluctuation correction, also called the *one-loop correction*, to the mean-field free energy in order to derive this quantity.

Subsequently we will present an alternative way to formulate a fluctuation-corrected Poisson-Boltzmann equation by means of a *variational method*. In the context of soft matter electrostatics, it has been originally proposed by Netz and Orland [4]. We subsequently apply this method to the adsorption of a weakly charged polymer to a rigid membrane [5].

2.1 The Partition Function of a Coulomb Gas of Interacting Ions

We want to address the statistical mechanics contained by the 'experimental' setup sketched in Fig. 2.1 that shows the interface of an *air-electrolyte system*. The model Hamiltonian for this system is given by

© The Author(s), under exclusive license to Springer Nature Switzerland AG 2023
R. Blossey, *The Poisson-Boltzmann Equation*,
SpringerBriefs in Physics, https://doi.org/10.1007/978-3-031-24782-8_2

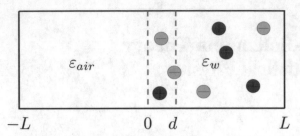

Fig. 2.1 A box of size $2L$ with the air phase and dielectric constant ε_{air} to the left and the water phase with the dielectric constant ε_w to the right. The interface between the two fluid phases is located at $z = 0$. In the interfacial water region $0 \leq z \leq d$ in the vicinity of the interface, the ions interact with a prescribed surface interaction potential $V_\pm(z)$ which models the specific interactions within the interfacial region

$$H = \frac{1}{2} \sum_{i \neq j} q_i q_j u(\mathbf{r}_i, \mathbf{r}_j) + \sum_i V_\pm(z_i) \tag{2.1}$$

where $q_i = \pm e$ is the charge of the monovalent ions. The first term in Eq. (2.1) is the *Coulomb interaction*

$$\nabla \cdot [\varepsilon(\mathbf{r}) \nabla u(\mathbf{r}, \mathbf{r}')] = -4\pi \delta(\mathbf{r} - \mathbf{r}'). \tag{2.2}$$

In Eq. (2.1), the summation runs over the indices $i \neq j$ which means that the (infinite) term corresponding to the self-interaction of the ions is subtracted out. The second term describes the ion-surface interaction in which anions and cations are treated separately. In order to limit the range of these interactions to within the interval $[0, d]$ we write explicitly

$$V_\pm(z_i) \equiv V_\pm(z_i)[\theta(z_i) - \theta(z_i - d)], \tag{2.3}$$

where $\theta(z)$ is the Heaviside function defined by

$$\theta(z) = \begin{cases} 0 & z \leq 0 \\ 1 & z \geq 0. \end{cases} \tag{2.4}$$

With this Hamiltonian we can write down the *grand partition function*

$$Z = \sum_{N_-=0}^{\infty} \sum_{N_+=0}^{\infty} \frac{\lambda_-^{N_-} \lambda_+^{N_+}}{N_-! \, N_+!} \int \prod_{i=1}^{N_-} d^3\mathbf{r}_i \prod_{j=1}^{N_+} d^3\mathbf{r}_j \exp(-\beta H). \tag{2.5}$$

The $N_\pm$ are the numbers of cations and anions. The *fugacities* $\lambda_\pm$ are given in terms of the chemical potentials $\mu_\pm$ via

$$\lambda_\pm = a^{-3} \exp(\beta \mu_\pm^{\text{tot}}) \tag{2.6}$$

with

$$\mu_\pm^{\text{tot}} = \mu_\pm(\mathbf{r}) + \frac{1}{2} u_{self}(\mathbf{r}, \mathbf{r}) \tag{2.7}$$

where $\mu_\pm^{\text{tot}}$ is the total chemical potential where the *self-energy* of the ions is subtracted. Further, the length a is a microscopic length, identified with the ion size; no difference is made between anions and cations.

Having defined the partition function, we take the next step to rewrite it as a field-theoretic expression. In order to achieve this we define the *charge density operator*

$$\widehat{\varrho}(\mathbf{r}) = \sum_j q_j \delta(\mathbf{r} - \mathbf{r}') \tag{2.8}$$

and the functional Dirac delta-function

$$\delta[\varrho(\mathbf{r}) - \widehat{\varrho}(\mathbf{r})] = \left(\frac{\beta}{2\pi}\right)^N \int \mathcal{D}\phi(\mathbf{r}) \exp\left[i\beta \int d^3\mathbf{r}\, \phi(\mathbf{r})\, [\varrho(\mathbf{r}) - \widehat{\varrho}(\mathbf{r})]\right] \tag{2.9}$$

in which $\phi(\mathbf{r})$ serves as an *auxiliary field*, and we have $N = N_+ + N_-$ as the total number of charges.

The insertion of this relationship in the partition function allows us to rewrite it as a functional over all values $\{\phi(\mathbf{r})\}$ with the help of the *Hubbard-Stratonovich transform*. For the one-dimensional case it reads as

$$\exp\left(-\frac{a}{2}x^2\right) = \frac{1}{\sqrt{2\pi a}} \int\limits_{-\infty}^{\infty} dy \exp\left(-\frac{1}{2a}y^2 - ixy\right). \tag{2.10}$$

For our partition function of interest we obtain therefore

$$Z = \frac{(2\pi)^{-N/2}}{\sqrt{\det[\beta^{-1} u(\mathbf{r}, \mathbf{r}')]}} \times \int \mathcal{D}\phi \exp\left[-\frac{\beta}{2} \int d^3\mathbf{r}\, d^3\mathbf{r}'\, \phi(\mathbf{r}) u^{-1}(\mathbf{r}, \mathbf{r}') \phi(\mathbf{r}')\right.$$

$$\left. + \int d^2\mathbf{r} \int\limits_0^d dz \left(\sum_{k=\pm} \lambda_k e^{ki\beta e\phi(\mathbf{r}) - \beta V_k(z)}\right) + \int d^2\mathbf{r} \int\limits_d^L \left(\sum_{k=\pm} \lambda_k e^{ki\beta e\phi(\mathbf{r})}\right)\right]. \tag{2.11}$$

Here,

$$u^{-1}(\mathbf{r}, \mathbf{r}') = -\frac{1}{4\pi} \nabla \cdot \left[\varepsilon(\mathbf{r}) \nabla \delta(\mathbf{r} - \mathbf{r}') \right] \tag{2.12}$$

is the inverse Coulomb potential which fulfills the integral relation

$$\int d^3\mathbf{r}'' u(\mathbf{r}, \mathbf{r}'') u^{-1}(\mathbf{r}, \mathbf{r}'') = \delta(\mathbf{r} - \mathbf{r}'). \tag{2.13}$$

The requirement of *electroneutrality* $e(\lambda_+ - \lambda_-) = 0$ means $\lambda_+ = \lambda_- = \lambda$ allows us to reexpress the partition function as

$$Z = \frac{(2\pi)^{-N/2}}{\sqrt{\det[\beta^{-1} u(\mathbf{r}, \mathbf{r}')]}} \int \mathcal{D}\phi \, e^{-H[\phi(\mathbf{r})]} \tag{2.14}$$

where

$$H[\phi(\mathbf{r})] = \int d^3\mathbf{r} \frac{\beta \varepsilon(\mathbf{r})}{8\pi} [\nabla \phi(\mathbf{r})]^2 + 2\lambda \int d^2\mathbf{r} \int_d^L dz \cos(\beta e \phi(\mathbf{r}))$$

$$-\lambda \int d^2\mathbf{r} \int_0^d dz \left(e^{-i\beta e\phi(\mathbf{r}) - \beta V_+(z)} + e^{i\beta e\phi(\mathbf{r}) - \beta V_-(z)} \right). \tag{2.15}$$

We further simplify by considering

$$\int d^2\mathbf{r} \int_0^d dz e^{-i\beta e\phi(\mathbf{r})} e^{-\beta V_+(z)} \approx \int d^2\mathbf{r} \left[\langle e^{-\beta V_+(z)} \rangle \int_0^d dz e^{-i\beta e\phi(\mathbf{r})} \right]$$

$$\equiv \int d^2\mathbf{r} \int_0^d dz e^{-i\beta e\phi(\mathbf{r}) - \beta\alpha_+}, \tag{2.16}$$

and likewise for the term involving $V_-(z)$. This step assumes that the details of the variation of ion concentrations in the layer $[0, d]$ can be neglected.

Our task now is to compute Z—which in fact is just too much to achieve. We will instead approximate this integral by the value at its extremum and consider also the quadratic variation around it. This so-called *saddle-point approximation* amounts to compute

$$H[\phi(\mathbf{r})] \approx H[\phi_0(\mathbf{r})] + \frac{1}{2} \int d^3\mathbf{r} d^3\mathbf{r}' \left. \frac{\delta^2 H[\phi(\mathbf{r})]}{\delta\phi(\mathbf{r})\delta\phi(\mathbf{r}')} \right|_0 \delta\phi(\mathbf{r})\delta\phi(\mathbf{r}'). \tag{2.17}$$

The second variation of the functional H under the integral is called the *Hessian*, and will be denoted by $H_2(\mathbf{r}, \mathbf{r}')$; its value is to be taken at the *stationary point* $\delta H/\delta\phi(\mathbf{r})|_0 = 0$ that corresponds to the solution of the Poisson-Boltzmann equation for our problem. With this quadratic approximation, the partition function is turned into a Gaussian integral which can be evaluated explicitly. From it results the *grand potential* $\Omega \equiv -k_B T \ln Z$ which reads as

$$\Omega \approx \Omega_0 + \Omega_1 = k_B T \left(S_0 + \frac{1}{2} \mathrm{Tr} \ln(H_2(\mathbf{r}, \mathbf{r}')) \right). \tag{2.18}$$

In this expression, constant terms have been dropped and the matrix identity

$$\ln \det(A) = \mathrm{Tr} \ln(A) \tag{2.19}$$

has been used. It follows from writing the matrix A in diagonal form. Because of this identity, the correction term to the saddle-point is also termed the *fluctuation determinant*.

2.2 Mean-Field Theory

We now solve the mean-field equations. Defining $\psi(\mathbf{r}) \equiv i\phi(\mathbf{r})$, the Poisson-Boltzmann equation obtained at the saddle-point implies three functions for the three regions shown in Fig. 2.1:

$$\nabla^2 \psi_1 = 0, \quad z < 0,$$

$$\nabla^2 \psi_2 = \frac{4\pi e n_b}{\varepsilon_w} \left(e^{-\beta\alpha_-} e^{\beta e \psi_2} - e^{-\beta\alpha_+} e^{-\beta e \psi_2} \right), \quad 0 \leq z \leq d,$$

$$\nabla^2 \psi_3 = \frac{8\pi e n_b}{\varepsilon_w} \sinh(\beta e \psi_3), \quad z > d. \tag{2.20}$$

The last equation corresponds to the Poisson-Boltzmann equation we discussed in detail in the first chapter. As was the case there, we have simplified the discussion to the consideration of only one dimension, taken as the z-direction, since we have translational symmetry in the transverse directions x and y. Furthermore, we will linearize the equations to go straight away to the Debye-Hückel limit. As before, this step requires $|\beta e \psi| \ll 1$. In this limit we have

$$\frac{d^2 \psi_1}{dz^2} = 0, \quad z < 0,$$

$$\frac{d^2 \psi_2}{dz^2} = \frac{1}{\beta e} \frac{\kappa_D^2}{2} \left(e^{-\beta\alpha_-} - e^{-\beta\alpha_+} \right) + \xi^2 \psi_2, \quad 0 \leq z \leq d,$$

$$\frac{d^2 \psi_3}{dz^2} = \kappa_D^2 \psi_3, \quad z > d. \tag{2.21}$$

where

$$\xi^2 \equiv ((\kappa_D^2)/2)(e^{-\beta\alpha_-} + e^{\beta\alpha_+}) \tag{2.22}$$

with

$$\kappa_D^2 = \frac{8\pi\beta e^2 n_b}{\varepsilon_w}. \tag{2.23}$$

Let us now solve these equations with some physical understanding. We need to exploit the boundary conditions which in this case mean that the potentials and their first derivatives at $z = 0$ and $z = d$ match. Since there is no electric field in the air phase $z < 0$ we immediately have

$$\psi_1 = A \tag{2.24}$$

with the constant A to be determined. For $z \to L$ in the bulk water phase, we know that the potential has to decay exponentially—we are in the Debye-Hückel situation. Thus

$$\psi_3(z) = De^{-\kappa_D z} \tag{2.25}$$

since the decay will happen on the scale of the Debye length. D is another constant to be found. In the intermediate region, we can write

$$\psi_2(z) = B + C\cosh(\alpha z). \tag{2.26}$$

We can have a constant contribution here, as this will drop out in the derivative. And for the hyperbolic function, a sinh is also not possible, as is would contribute in the first derivative.

Now, let's check. From the condition $\psi_1(0) = \psi_2(0)$ we find

$$A = B + C. \tag{2.27}$$

The condition $\psi_1'(0) = \psi_2'(0)$ is immediately fulfilled. (Note: $\psi' = d\psi/dz$). At the right end of the interval, $z = d$, we have $\psi_2(d) = \psi_3(d)$, thus

$$B = C\cosh(\alpha d). \tag{2.28}$$

And finally, for the derivatives $\psi_2'(d) = \psi_3'(d)$ we find

$$C\alpha\sinh(\alpha d) = -\kappa_D De^{-\kappa_D d}. \tag{2.29}$$

In order to identify the constants, we can start with the equation for $\psi_2(z)$. For our solution ansatz we have

$$\psi_2(z) = C\alpha^2\cosh(\alpha z) = \alpha^2(\psi_2(z) - B). \tag{2.30}$$

We can therefore identify

$$\alpha = \xi \tag{2.31}$$

and

$$B = -\frac{1}{\xi^2} \frac{1}{\beta e} \frac{\kappa_D^2}{2} \left(e^{-\beta\alpha_-} - e^{-\beta\alpha_+} \right). \tag{2.32}$$

Hence, we can compute C from Eq. (2.29), and A and D follow as well. The final result reads as the following somewhat lengthy expressions

$$\psi_1 = \frac{\kappa_D(\cosh(\xi d) - 1) + \xi \sinh(\xi d)}{\kappa_D \cosh(\xi d) + \xi \sinh(\xi d)} \chi$$

$$\psi_2(z) = \frac{\kappa_D(\cosh(\xi d) - \cosh(\xi z)) + \xi \sinh(\xi d)}{\kappa_D \cosh(\xi d) + \xi \sinh(\xi d)} \chi$$

$$\psi_3(z) = \frac{\xi \sinh(\xi d)}{\kappa_D \cosh(\xi d) + \xi \sinh(\xi d)} e^{-\kappa_D(z-d)} \chi. \tag{2.33}$$

The parameter χ is given by

$$\chi \equiv \frac{1}{\beta e} \frac{e^{-\beta\alpha_+} - e^{-\beta\alpha_-}}{e^{-\beta\alpha_+} + e^{-\beta\alpha_-}}. \tag{2.34}$$

If one considers the intermediate region as very small, one can expand the functions to linear order which leads to

$$\psi_1 = \frac{\kappa_D d}{2\beta e} (e^{-\beta\alpha_+} - e^{-\beta\alpha_-})$$

$$\psi_2 = \psi_1$$

$$\psi_3(z) = \psi_1 e^{-\kappa_D(z-d)} \chi. \tag{2.35}$$

If one finally sets $\alpha_+ = 0$, one arrives at the case of a single type of ion being subjected to the surface interaction, and this case is equivalent to the limit of a proximal layer of zero width. This is the simplified case we will now address for the determination of the one-loop correction.

2.3 One-Loop Correction

Since we want to find the one-loop correction, the next task is the computation of the fluctuation determinant, Eq. (2.18). This requires the development of some technique, and this is better done by considering a simplified case first. The simplification we introduce is to shrink the interfacial region to zero, $d \to 0$, and to replace the charges contained in the intermediate region with a surface charge localized at $z = 0$. Our previous region 3 then will be called region 2, while the earlier Sect. 2.2 has shrunk to zero. Our problem thus reduces to [2]

$$\nabla^2 \psi_1 = 0, \quad z < 0,$$
$$\nabla^2 \psi_2 = \frac{8\pi e n_b}{\varepsilon_w} \sinh(\beta e \psi_2), \quad z > d. \tag{2.36}$$

while at $z = 0$ we have the boundary condition

$$\varepsilon_w \left. \frac{\psi_2}{dz} \right|_{0+} - \varepsilon_a \left. \frac{d\psi_1}{dz} \right|_{0-} = -4\pi \sigma_0 e^{\beta e \psi_0} \tag{2.37}$$

where $\sigma_0 e^{\beta e \psi_0}$ is the effective surface charge with

$$\sigma_0 = -a e n_b (e^{-\beta \alpha} - 1). \tag{2.38}$$

The two parameters $\alpha_\pm$ are thus reduced to a single parameter, which for $\alpha > 0$ renders $\sigma_0 > 0$. Linearizing the Poisson-Boltzmann equation in region 2 again, as before, we find the solutions in the air and water phases to be

$$\nabla^2 \psi_1(z) = \psi_0 = \frac{4\pi \sigma_0}{\kappa_D \varepsilon_w - 4\pi \beta e \sigma_0}$$

$$\nabla^2 \psi_2(z) = \psi_0 e^{-\kappa_D z} \quad z > 0. \tag{2.39}$$

Having simplified the zeroth-order problem, in the next step we express the one-loop contribution to the grand potential in terms of the *secular determinant*

$$\Omega_1 = \frac{A k_B T}{8\pi^2} \int d^2 k \ln \left(\frac{D_{\nu=1}(k)}{D_{\nu=0}(k)} \right) \tag{2.40}$$

where the integral is taken over the transverse wave-vector $\mathbf{k} = (k_x, k_y)$ and the index ν refers to the differential operator in the definition of D_ν given by

$$D_\nu \equiv \det \left[\frac{d^2}{dz^2} - k^2 - \nu \kappa_D^2 \cosh(\beta e \psi_0 e^{-\kappa_D z}) \right], \tag{2.41}$$

i.e. defined for the initial value problem

$$\frac{d^2 f_\nu}{dz^2} - (k^2 + \nu \kappa_D^2 \cosh(\beta e \psi_0 e^{-\kappa_D z})) f_\nu(z) = 0, \tag{2.42}$$

with $f_v(z \to \pm\infty) = 0$ and with the boundary condition at $z = 0$ given by

$$\varepsilon_w \frac{df_v}{dz}(0+) - \varepsilon_a \frac{df_v}{dz}(0-) = \omega_v f_v(0) \tag{2.43}$$

where $\omega_v = -4\pi v \beta e \sigma_0 e^{\beta a \psi_0}$.

At this point we need to make a little break in this discussion in order to explain how we are actually practically calculating the expression (2.40). Let us discuss this at an even simpler example. For this we take the differential operator, following Ref. [6]:

$$L_1 = \left(-\frac{d^2}{dx^2} + m^2 \right). \tag{2.44}$$

The eigenvalue spectrum of this operator acting on the function $\psi(x)$ fulfills

$$L_1 \psi_n(x) = \lambda_n \psi_n(x) \tag{2.45}$$

with the boundary conditions $\psi_n(0) = 0$, $\psi'_n(0) = 1$. We consider solutions on the interval $[0, L]$. The eigenvalues are then given by

$$\lambda_n = m^2 + \left(\frac{n\pi}{L} \right)^2. \tag{2.46}$$

If we now look at the determinant ratio of the two differential operators L_1 and $L_2 = -d^2/dx^2$ we find

$$\frac{\det L_1}{\det L_2} = \prod_{n=1}^{\infty} \frac{\left(m^2 + \left(\frac{n\pi}{L} \right)^2 \right)}{\left(\frac{n\pi}{L} \right)^2} = \prod_{n=1}^{\infty} \left(1 + \left(\frac{mL}{n\pi} \right)^2 \right) = \frac{\sinh(mL)}{mL}. \tag{2.47}$$

Obviously, this approach requires the knowledge of the full eigenvalue spectrum of L_1 which in this case is easy to find. For our case, this is not so. There is, however, a much more elegant approach based on a *theorem of Gel'fand and Yaglom* [6]. It just requires to solve two ordinary differential equations on the interval $[0, L]$. They are

$$L_1 \phi(x) = 0, \quad \phi(x) = \frac{\sinh(mL)}{mL} \tag{2.48}$$

and

$$L_2 \phi_0(x) = 0, \quad \phi_0(x) = x. \tag{2.49}$$

The determinant ration is then given simply by

$$\frac{\det L_1}{\det L_2} = \frac{\phi(L)}{\phi_0(L)} = \frac{\sinh(mL)}{mL}. \tag{2.50}$$

We have thus obtained the same result.

In our case, the boundary conditions are more complicated, but the Gel'fand-Yaglom theorem also covers these. Linearizing Eq. (2.42) we can write the equation for f_ν as

$$\left[\frac{d^2}{dz^2} - p_\nu^2\right] f_\nu(z) = 0 \tag{2.51}$$

with $p_\nu^2 \equiv k^2 + \nu\kappa_D^2$. The general solution of this equation is

$$\begin{aligned} f_1(z) &= A_1 e^{kz}, & z &< 0 \\ f_2(z) &= A_2 C_\nu(z) + B_2 S_\nu(z), & z &> 0. \end{aligned} \tag{2.52}$$

The functions $C_\nu(z)$ and $S_\nu(z)$ are given by

$$C_\nu \equiv \cosh(p_\nu z), \quad S_\nu(z) \equiv \frac{\sinh(p_\nu z)}{p_\nu}. \tag{2.53}$$

hence the even and odd solutions for $z > 0$ fulfilling the boundary conditions

$$\begin{aligned} C_\nu(0) &= 1, & C_\nu'(0) &= 0 \\ S_\nu(0) &= 0, & S_\nu'(0) &= 1. \end{aligned} \tag{2.54}$$

The $'$-symbol stands again for the derivative d/dz.

In the generalized Gel'fand-Yaglom approach we can now combine the boundary conditions by considering the problem also on the finite interval $[0, L]$:

$$D_\nu = \det\left[M + N\begin{pmatrix} C_\nu(L) & S_\nu(L) \\ C_\nu'(L) & S_\nu'(L) \end{pmatrix}\right], \tag{2.55}$$

where the matrices M and N satisfy the boundary-value equation

$$M\begin{pmatrix} u_\nu(0) \\ u_\nu'(0) \end{pmatrix} + N\begin{pmatrix} u_\nu(L) \\ u_\nu'(L) \end{pmatrix} = 0 \tag{2.56}$$

for $u_\nu = C_\nu$ or $u_\nu = S_\nu$. The matrices M and N can be chosen as

$$M = \begin{pmatrix} -(\omega_\nu + \varepsilon_a k) & \varepsilon_w \\ 0 & 0 \end{pmatrix}, \quad N = \begin{pmatrix} 0 & 0 \\ 0 & 1 \end{pmatrix}. \tag{2.57}$$

The resulting secular determinant reads as

$$D_\nu = -(\omega_\nu + \varepsilon_a k)\cosh(p_\nu L) - p_\nu\varepsilon_w\sinh(p_\nu L). \tag{2.58}$$

In the limit $L \to \infty$ we have

$$D_v = -\frac{1}{2} \left[\omega_v + \varepsilon_a k + \varepsilon_w p_v \right] e^{p_v L}, \qquad (2.59)$$

and thus

$$\frac{D_1}{D_0} = \frac{\omega + \varepsilon_a k + \varepsilon_w p}{(\varepsilon_w + \varepsilon_a)k} e^{(p-k)L}, \qquad (2.60)$$

where $\omega \equiv \omega_1$ and $p \equiv p_1$. One thus obtains for Ω_1 the final result

$$\Omega_1 = \frac{V k_B T}{12\pi} \left[(\Lambda^2 + \kappa_D^2)^{3/2} - \Lambda^3 - \kappa_D^3 \right] + \frac{A k_B T}{4\pi} \int_0^\Lambda dk k \ln \left(\frac{\omega + \varepsilon_a k + \varepsilon_w p}{(\varepsilon_w + \varepsilon_a)k} \right), \qquad (2.61)$$

where the integrals are defined with a so-called *UV-cutoff* to control formal divergences of the mathematical expressions at microscopic scales. We take $\Lambda \equiv 2\sqrt{\pi}/a$ with a as the average minimal distance between the ions.[1]

What does the result look like in the case of a finite region of width d at the waterside of the air-water interface? The calculation above carries essentially through by now taking into account the boundary conditions at $z = 0$, $z = d$ and $z = L$, and reads as

$$D(k) \approx e^{pL} e^{(p-q)d} \left(\frac{\varepsilon_w q + \varepsilon_a k}{q} \right) \left[q(1 - \Delta(q,k)e^{-2qd}) + p(1 + \Delta(q,k)e^{-2qd}) \right] \qquad (2.62)$$

with

$$q^2 = k^2 + \xi^2, \qquad (2.63)$$

and

$$\Delta(q,k) \equiv \frac{\varepsilon_w q - \varepsilon_a k}{\varepsilon_w q + \varepsilon_a k}. \qquad (2.64)$$

Keeping only linear terms in d, we find

$$D(k) \approx \left[\varepsilon_a k + \varepsilon_w p + \varepsilon_w d(\xi^2 - \kappa_D^2) \right] e^{pL}, \qquad (2.65)$$

---

[1] Yes, a is used a lot in this book with many different meanings, but they should be clear from the context.

which shows how the presence of the intermediate region generalizes in comparison to the simplified model.

From these results we can write down the *grand potential* Ω for the three-segment configuration of Fig. 2.1 up to one-loop order as

$$
\begin{aligned}
\Omega = \Omega_0 &+ \frac{V k_b T}{12\pi} \left[(\Lambda^2 + \kappa_D^2)^{3/2} - \kappa_D^3 - \Lambda^3 \right] \\
&+ \frac{A d k_B T}{12\pi} \left[(\Lambda^2 + \xi^2)^{3/2} - (\Lambda^2 + \kappa_D^2)^{3/2} - \xi^3 + \kappa_D^3 \right] \quad (2.66) \\
&+ \frac{A k_B T}{4\pi} \int_0^\Lambda dk\, k \left(\ln \left[\frac{\varepsilon_w q + \varepsilon_a k}{2(\varepsilon_w + \varepsilon_a) k q} \right] \right. \\
&\left. + \ln[q(1 - \Delta(q,k) e^{-2qd}) + (1 + \Delta(q,k) e^{-2qd})] \right).
\end{aligned}
$$

This result ends this section; we can now move on to compute the surface tension.

2.4 The Free Energy and the Surface Tension

The *surface tension* is defined in terms of the difference between the free energies of the given system and the corresponding homogeneous bulk phases via the definition

$$
\Delta\gamma \equiv \left[F(2L) - F^{(water)}(L) - F^{(air)}(L) \right] / A \quad (2.67)
$$

where the contributions of two homogenous slabs of length L of water and air phases are subtracted from the full free energy. Thus, we need to first write down these expressions, starting from the grand potential Ω.

The relation between the free energy F and Ω is given by the expression

$$
F = \Omega + \sum_{i=\pm} \int d^3\mathbf{r}\, \mu_i(\mathbf{r}) n_i(\mathbf{r}). \quad (2.68)
$$

Since the grand potential is expressed in terms of the fugacities $\lambda_\pm$ for anions and cations, we first need to find the proper expression of Ω in terms of the ion density. The relation is given by

$$
n_b^{(i)} = -\lambda_i \frac{\beta}{V} \frac{\partial \Omega}{\partial \lambda_i}, \quad (2.69)
$$

where Ω is the free energy to all loop orders. To the order of one loop, the one we are interested in, we write

$$
F = \Omega_0(\lambda_i) + \Omega_1(\lambda_i) + k_B T V \sum_{i=\pm} n_b^{(i)} \ln(\lambda_i a^3) - \frac{1}{2} e^2 n_b^{(i)} u_{self}(\mathbf{r}, \mathbf{r}) \quad (2.70)
$$

Assuming now $\lambda_i = \lambda_{i,0} + \lambda_{i,1}$ we can expand all λ_i-dependent terms and obtain to lowest order in the corresponding terms

$$\Omega_0(\lambda_i) = \Omega(\lambda_{i,0}), \quad \Omega_1(\lambda_i) = \Omega_1(\lambda_{i,0}) + \lambda_{i,1} \frac{\partial \Omega_1}{\partial \lambda_i}\bigg|_{\lambda_i = \lambda_{i,0}} \tag{2.71}$$

and for the logarithmic term we have

$$\ln(\lambda_i a^3) = \ln((\lambda_{i,0} + \lambda_{i,1})a^3) = \ln\left(\lambda_{i,0} a^3 \left(1 + \frac{\lambda_{i,1}}{\lambda_{i,0}}\right)\right)$$

$$\approx \ln(\lambda_{i,0} a^3) + \frac{\lambda_{i,1}}{\lambda_{i,0}}. \tag{2.72}$$

With the help of relation (2.69) we see that the $\lambda_{i,1}$-dependent terms cancel out and we are left with an expression solely in $\lambda_{0,i}$ and we therefore can replace the fugacity by ion density up to one-loop order.

Before we will embark on the surface tension, we will first write down the expressions for the bulk and surface contributions of the free energy of our system, i.e., we compute F/V and F/A up to one loop order. We will do so in the limit $\Lambda \to \infty$, starting from Eq. (2.66). The result is

$$\frac{F}{V} \approx \frac{\Omega_0}{V} + 2k_B T n_b \ln(n_b a^3) - \frac{k_B T}{12\pi} \kappa_D^3 - \frac{d}{L} \frac{k_B T}{12\pi} (\xi^3 - \kappa_D^3) \tag{2.73}$$

$$+ \frac{k_B T}{8\pi} \Lambda \left[\kappa_D^2 + \frac{d}{L}(\xi^2 - \kappa_D^2)\right] - e^2 n_b u_{self}(\mathbf{r}, \mathbf{r})\left(1 + \frac{d}{L} \frac{\xi^2 - \kappa_D^2}{\kappa_D^2}\right).$$

The last term merits some further discussion. First of all, the factor

$$n_b \left(1 + \frac{d}{L} \frac{\xi^2 - \kappa_D^2}{\kappa_D^2}\right) \tag{2.74}$$

corresponds to the density term $n_b V$, taking into account the finite intermediate region of length d. To see this we write

$$n_b V = n_b(A(L - d) + Ad). \tag{2.75}$$

The ion density is different in the intermediate and the right compartment. We thus write

$$n_b = n_b\left(1 - \frac{d}{L}\right)\frac{\kappa_D^2}{\kappa_D^2} + \frac{d}{L}\frac{\xi^2}{\kappa_D^2} = n_b\left(1 + \frac{d}{L}\frac{\xi^2 - \kappa_D^2}{\kappa_D^2}\right). \tag{2.76}$$

Even more importantly, we have to see what happens to the self-energy of the ions $u(\mathbf{r}, \mathbf{r})$, as this term is formally divergent. So we need to take a closer look and properly compute it. It is obtained from the Coulomb potential which fulfills

$$\nabla^2 u(\mathbf{r}, \mathbf{r}') = -\frac{4\pi}{\varepsilon_w} \delta(\mathbf{r} - \mathbf{r}'), \quad z \geq 0$$

$$\nabla^2 u(\mathbf{r}, \mathbf{r}') = 0, \qquad\qquad z < 0. \tag{2.77}$$

Employing the translational invariance of the system in transverse directions we can simplify with the help of the *Fourier-Bessel representation* to obtain

$$u(\mathbf{r}, \mathbf{r}') = \frac{1}{4\pi^2} \int d^2\mathbf{k}\, U_0(k; z, z')\, e^{i\mathbf{k}\cdot\varrho} = \int dk k\, U_0(k; z, z')\, J_0(k, |\varrho - \varrho'|). \tag{2.78}$$

where J_0 is the *zeroth-order Bessel function of the first kind*, and ϱ the transverse, radial component of the spatial vector. The solution of Eq. (2.77) for the positive half-space $z \geq 0$ is given by

$$U_0(k; z, z') = \frac{2\pi}{\varepsilon_w k} \left(e^{-k|z-z'|} + \frac{\varepsilon_w - \varepsilon_a}{\varepsilon_w + \varepsilon_a} e^{-k(z+z')} \right) \tag{2.79}$$

while for $z \leq 0$ one finds

$$U_0(k; z, z') = \frac{2\pi}{\varepsilon_w k} \left(1 + \frac{\varepsilon_w - \varepsilon_a}{\varepsilon_w + \varepsilon_a} e^{k(z-z')} \right). \tag{2.80}$$

The self-energy of an ion in the bulk then is found by putting $\mathbf{r} = \mathbf{r}'$ and considering $z \to \infty$ so that

$$u_{self}(\mathbf{r}, \mathbf{r}) = \frac{1}{\varepsilon_w} \int_0^{\Lambda} dk \left(1 + \frac{\varepsilon_w - \varepsilon_a}{\varepsilon_w + \varepsilon_a} e^{-2kz} \right) \approx \frac{\Lambda}{\varepsilon_w}. \tag{2.81}$$

for large z. As a consequence, the two last terms in F/V cancel each other exactly.

The contribution of the surface free energy to one-loop order is given by

$$\frac{F}{A} = \frac{k_B T}{4\pi} \int_0^{\Lambda} dk k \mathcal{I}[k] \tag{2.82}$$

where

$$\mathcal{I}[k] \equiv \ln\left[\frac{\varepsilon_w q + \varepsilon_a k}{2(\varepsilon_w + \varepsilon_a)kq} \right] + \ln\left[q(1 - \Delta(q, k)e^{-2qd}) + p(1 + \Delta(q, k)e^{-2qd}) \right]. \tag{2.83}$$

We now have assembled all information that is needed to compute the surface tension according to Eq. (2.67). In that expression, the contribution $F^{(air)}(L) = 0$, since there are no ions in the air phase. The contribution $F^{(water)}(L)$ can be computed from the knowledge of $F(2L)$ by considering the interval $[-L, L]$ as a uniform dielectric medium with dielectric constant ε_w and $\alpha_\pm = 0$, $q = p$. One thus has $D(\mathbf{k})/D_{free}(\mathbf{k}) = p/k$, and $\Omega_0 = -2n_B V$ for the bulk water phase in which $\psi = 0$. Taking into account the factor of two in the interval size, we have

$$F^{(water)}(L) = k_B T V \left[-2n_b + 2n_B \ln(n_b a^3) - \frac{\kappa_D^3}{12\pi} \right] + \frac{k_B T A}{8\pi} \int_0^\Lambda dk k \ln \frac{p}{k}.$$

(2.84)

The surface tension can then be written as the sum of the mean-field and one-loop contributions in the form

$$\Delta \gamma = \Delta \gamma_0 + \Delta \gamma_1$$

(2.85)

with

$$\Delta \gamma_0 = -\int_{-\infty}^\infty dz \frac{\varepsilon(z)}{8\pi} \left(\frac{d\psi}{dz} \right)^2$$

$$+2L k_B T n_b - \int_0^d dz (e^{-\beta e\psi} e^{-\beta\alpha_+} - e^{\beta e\psi} e^{-\beta\alpha_-})$$

$$-2k_B T n_b \int_d^L dz \cosh(\beta e\psi)$$

(2.86)

where the mean-field solution for the electrostatic potential, $\psi(z)$, enters. This expression can be rewritten as, see Ref. [3]

$$\Delta \gamma_0 = -k_B T n_b d \left[(e^{-\beta\alpha_-} - 1) - (e^{\beta\alpha_+} - 1) \right].$$

(2.87)

Finally, the result for the one-loop contribution reads as

$$\Delta \gamma_1 = \frac{k_R T}{8\pi} \int_0^\Lambda dk k \left(\ln \left[\frac{k}{p} \left(\frac{\varepsilon_w q + \varepsilon_a k}{2(\varepsilon_w + \varepsilon_a)kq} \right)^2 \right] \right.$$

$$+ \ln \left[q(1 - \Delta(q,k)e^{-2qd}) + p(1 + \Delta(q,k)e^{-2qd}) \right]^2 \bigg)$$

$$- \frac{2d}{3}(\xi^3 - \kappa_D^3),$$

(2.88)

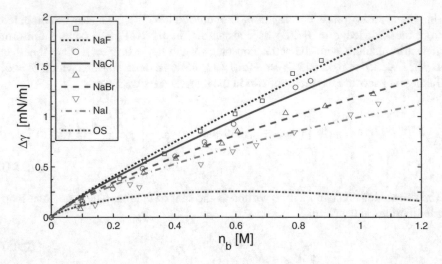

Fig. 2.2 Comparison of the fitted excess surface tension $\Delta\gamma$ at the air/water interface with experimental data for monovalent salts from the Na^+ series; Figure taken from [3]. Reprinted with permission from EPL

where $q^2 = k^2 + \xi^2$ and $\Delta(q, k)$ is given by Eq. (2.64).

We end this section with a brief comparison with experiments. Markovich, Andelman and Podgornik in their work [2, 3] provide a detailed discussion of the comparison of their theoretical results described in this section with a number of experiments taken from the literature. We do not reproduce this whole discussion, but do show one exemplary result. Figure 2.2 show the fits of the excess surface $\Delta\gamma$ at the air/water interface with experimental data for the Na^+ series of salts as a function of salt number density. The α-parameters were chosen as fit parameters for these curves. The data very clearly follow the trends given by the theoretical curves, and are far superior than the early results from Onsager-Samaras [1], indicated by 'OS' in the figure.

2.5 The Variational Method

We continue our discussion of statistical mechanics approaches to PB-theory in this section by discussing an alternative method to obtain a fluctuation-corrected Poisson-Boltzmann theory. It consists of the application of a variational method in order to obtain first an approximation to the partition function by invoking a suitable approximate Hamiltonian, and was first proposed in the context of Poisson-Boltzmann in Ref. [4]. We follow their path, as well as later work [7].

Starting point of our calculations is again the (grand canonical) partition function for our standard PB-theory of (1:1) salt

$$Z = \int \mathcal{D}\phi \, e^{-H[\phi]} \tag{2.89}$$

with the Hamiltonian functional $H[\phi]$ now given by

$$H[\phi] = \int \frac{d\mathbf{r}}{2\pi} \left[\frac{(\nabla \phi(\mathbf{r}))^2}{4} + i\phi(\mathbf{r})\varrho_f - \frac{\Lambda}{2} e^{\frac{\Xi}{2} G_0(\mathbf{r},\mathbf{r})} \cos \phi(\mathbf{r}) \right]. \tag{2.90}$$

Two rescalings have been undertaken: first, the field has been rescaled via $e\beta\phi \to \phi$ and then space in terms of the Gouy-Chapman length ℓ_{GC}, and charge density and fugacity been redefined accordingly; for Λ we have $\Lambda = 8\pi\lambda\ell_{GC}^3 \Xi$ with Ξ defined by

$$\Xi \equiv q^2 \frac{\ell_B}{\ell_{GC}}, \tag{2.91}$$

which is called the *coupling parameter*; λ relates to the chemical potential. The value of the parameter Ξ can be used to delineate different fluctuation regimes, a weak, an intermediate and a strong regime; we do not get into this discussion here; see the suggestions for further reading at the end of the chapter.

For our purpose here, the main new feature in this expression is the exponential function, although in fact we encountered something similar in passing already in the previous section; here we make the discussion explicit. The term alluded to is

$$\sim e^{\frac{\Xi}{2} G_0(\mathbf{r},\mathbf{r})}$$

in the prefactor of the ion-term in the PB-theory. G_0 is the bare Coulomb interaction given by

$$G_0(\mathbf{r}, \mathbf{r}') = \frac{1}{|\mathbf{r} - \mathbf{r}'|} \tag{2.92}$$

which obviously diverges in the limit $\mathbf{r} \to \mathbf{r}'$. We here are again confronted with the self-energy problem, i.e. the singular behaviour of Coulomb forces when coordinates coincide. Here, the nominally singular prefactor has been introduced to take care of a singular term arising in the variational method, as we will now see.

The variational method consists in computing the free energy (Gibbs or other) from the expression

$$F = F_0 + \langle H - H_0 \rangle_0 / \Xi \tag{2.93}$$

with the reference or trial Hamiltonian H_0 of the most general quadratic form given by

$$H_0[\phi] = \frac{1}{2} \int d\mathbf{r} d\mathbf{r}' \, [\phi + i\Phi_0]_{\mathbf{r}} \, (\Xi G)^{-1}(\mathbf{r}, \mathbf{r}') \, [\phi + i\Phi_0]_{\mathbf{r}'} . \tag{2.94}$$

The field-theoretic average of a general functional $M[\phi]$ with respect to the reference Hamiltonian is defined as

$$\langle M[\phi]\rangle_0 \equiv \frac{\int \mathcal{D}\phi \; e^{-H_0[\phi]} M[\phi]}{\int \mathcal{D}\phi \; e^{-H_0[\phi]}}. \tag{2.95}$$

In our case, the trial functions are given by the mean electrostatic potential $\Phi(\mathbf{r})$ and the covariance G. For our model Hamiltonian of Eq. (2.89) of the 'modified' standard PB-equation, the free energy reads as

$$F = -\frac{1}{2}\text{Tr}\ln(\Xi G) - \int d\mathbf{r}\left[\frac{(\nabla\Phi)^2}{8\pi\Xi} - \frac{\varrho_f\Phi}{2\pi\Xi}\right]$$
$$+ \frac{1}{8\pi}\int d\mathbf{r}\int d\mathbf{r}'\delta(\mathbf{r}-\mathbf{r}')\nabla_{\mathbf{r}}\nabla_{\mathbf{r}'}G(\mathbf{r},\mathbf{r}')$$
$$- \frac{\Lambda}{4\pi\Xi}\int d\mathbf{r}e^{\frac{\Xi}{2}[G(\mathbf{r},\mathbf{r}')-G_0(\mathbf{r},\mathbf{r})]}\cosh\Phi, \tag{2.96}$$

where the correlation function or self energy of a mobile ion is given by

$$C(\mathbf{r}) \equiv \lim_{\mathbf{r}'\to\mathbf{r}}\left[G(\mathbf{r},\mathbf{r}') - G_0(\mathbf{r},\mathbf{r}')\right]. \tag{2.97}$$

We see that in the limit $\mathbf{r}\to\mathbf{r}'$, the argument in the exponential function of the last term stays finite, and thus the final theory we wish to work with has no pathology.

The last remaining step now is the variation of the approximate variational functional with respect to the electrostatic field Φ and G. The resulting equations read as

$$\nabla^2\Phi(\mathbf{r}) - \Lambda e^{-\frac{\Xi}{2}C(\mathbf{r})}\sinh\Phi(\mathbf{r}) = -2\varrho_f \tag{2.98}$$

and

$$\left[\nabla^2\Phi(\mathbf{r}) - \Lambda e^{-\frac{\Xi}{2}C(\mathbf{r})}\cosh\Phi(\mathbf{r})\right]G(\mathbf{r},\mathbf{r}') = -4\pi\delta(\mathbf{r}-\mathbf{r}'). \tag{2.99}$$

Equation (2.99) is a modified Debye-Hückel equation with the squared inverse screening length given by

$$\kappa^2 \equiv \Lambda e^{\frac{-\Xi}{2}C(\mathbf{r})}. \tag{2.100}$$

An alternative method to derive these equations employs a standard field-theoretic protocol via the *Schwinger-Dyson equation* and the derivation of two *Ward identities* [8]. For this one considers the path integral

$$\int D[\phi]\frac{\delta}{\delta J}e^{-H[\phi]+\int d\mathbf{r}J(\mathbf{r})\phi(\mathbf{r})} = 0. \tag{2.101}$$

The two variational Eqs. (2.98) and (2.99) are then obtained by performing the functional derivative with respect to the external current J and setting it to $J = 0$

afterwards; the second variational equation follows from the equation

$$\frac{\delta}{\delta J(\mathbf{r}')}\left[\int D[\phi]\frac{\delta}{\delta J(\mathbf{r})}e^{-H[\phi]+\int d\mathbf{r}J(\mathbf{r})\phi(\mathbf{r})}\right]=0. \tag{2.102}$$

The resulting averages are then computed with the variational Hamiltonian.

2.6 A Charged Polymer Interacting with a Like-Charged Membrane

In this section we will discuss how the variational method can be applied in practice; more will come in Chap. 3. The problem we want to tackle is that of a charged polymer in the vicinity of a membrane. The polymer can be either attracted or repelled from the surface. Supposing that both have equal charges—amounting to a repulsion within mean-field theory—what role is played by fluctuation effects in this situation?

The setup is the configuration shown in Fig. 2.3, in which a *stiff polymer* of length L is shown parallel to a rigid membrane surface with dielectric constant ε_m. The starting point for the mathematical description of the problem is a specific form of the variational grand potential given by the expression [5]

$$\begin{aligned}\Omega_{var}=&-\frac{1}{2}\text{Tr}\ln[v]+\int d\mathbf{r}\sigma(\mathbf{r})\psi(\mathbf{r})\\&+\frac{k_BT}{2e^2}\int d\mathbf{r}\varepsilon(\mathbf{r})\left[\nabla_{\mathbf{r}'}\cdot\nabla_{\mathbf{r}}v(\mathbf{r},\mathbf{r}')|_{\mathbf{r}'\to\mathbf{r}}-(\nabla\psi(\mathbf{r}))^2\right]\\&-\sum_i\Lambda_i\int d\mathbf{r}e^{-V_i(\mathbf{r})}e^{-q_i\psi(\mathbf{r})}e^{-\frac{q_i^2}{2}v(\mathbf{r},\mathbf{r})}.\end{aligned} \tag{2.103}$$

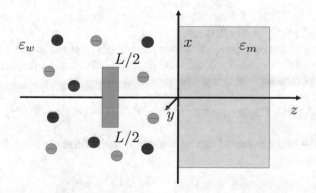

Fig. 2.3 System setup of the interaction of a charged polymer of length L in a solvent with permittivity ε_w with a membrane with permittivity ε_m. The polymer carries the total charge $Q_p = \tau L$, where τ is the line-charge density

In this expression, we use the symbol $v(\mathbf{r}, \mathbf{r}')$ for the variational Green's function, in accord with the use in [5]. $\psi(\mathbf{r})$ is the electrostatic potential induced by the density of fixed charges, $\sigma(\mathbf{r})$. The other parameters have been used before, $V_i(\mathbf{r})$ is the ionic steric potential due to the rigid membrane.

For our standard case of (1:1) salt we obtain the following variational equations:

$$\nabla \varepsilon(\mathbf{r}) \cdot \nabla \psi(\mathbf{r}) - \frac{2\varrho_b q e^2}{k_B T} e^{-V_i(\mathbf{r}) - \frac{q^2}{2} \delta v(\mathbf{r})} \sinh(q\psi(\mathbf{r}))$$

$$= -\frac{e^2}{k_B T} \sigma(\mathbf{r}), \tag{2.104}$$

and

$$\nabla \varepsilon(\mathbf{r}) \cdot \nabla v(\mathbf{r}, \mathbf{r}') - \frac{2\varrho_b q^2 e^2}{k_B T} e^{-V_i(\mathbf{r}) - \frac{q^2}{2} \delta v(\mathbf{r})} \cosh(q\psi(\mathbf{r})) v(\mathbf{r}, \mathbf{r}') = -\frac{e^2}{k_B T} \delta(\mathbf{r} - \mathbf{r}'). \tag{2.105}$$

The ionic self-energy $\delta v(\mathbf{r})$ is defined as

$$\delta v(\mathbf{r}) = \lim_{\mathbf{r} \to \mathbf{r}'} [v(\mathbf{r}, \mathbf{r}') - v_b(\mathbf{r} - \mathbf{r}')] \tag{2.106}$$

with the Debye-Hückel potential

$$v_b(\mathbf{r} - \mathbf{r}') = \ell_b \frac{e^{-\kappa_b |\mathbf{r} - \mathbf{r}'|}}{|\mathbf{r} - \mathbf{r}'|}. \tag{2.107}$$

The Debye screening parameter is given by $\kappa_b^2 = 8\pi q^2 \ell_B \varrho_b$ with the Bjerrum length $\ell_B = e^2/(4\pi \varepsilon_w k_B T)$, and

$$\varrho_b = \Lambda_i e^{-\frac{q_i^2}{2} v_b(0)} \tag{2.108}$$

relating ionic fugacity and bulk density.

A rescaling of the equations via $\phi \equiv q\psi$, $u \equiv q^2 v$ and the length $\mathbf{r} \to \kappa_b \mathbf{r}$ one can expand both the rescaled potential and the rescaled Green's function in terms of the coupling parameter $\Gamma = q^2 \kappa_b \ell_b$ as

$$\phi(\mathbf{r}) = \phi_0(\mathbf{r}) + \Gamma \phi_1(\mathbf{r}) + O(\Gamma^2), \quad u(\mathbf{r}, \mathbf{r}') = \Gamma u_1(\mathbf{r}, \mathbf{r}') + O(\Gamma^2) \tag{2.109}$$

from which the grand potential can be organized in the form

$$\Omega = \frac{1}{\Gamma} \Omega_{MF} + \Omega_u + \Gamma \Omega_{(\phi_1)}. \tag{2.110}$$

Here we will only take into account the lowest order contribution.

So far we have only dealt with the electrolyte-membrane system: we need to put in the charged polymer. The total charge density is composed by the sum of the charge densities of the membrane and the polymer, hence we have

$$\sigma(\mathbf{r}) = \sigma_m(\mathbf{r}) + \sigma_p(\mathbf{r}). \tag{2.111}$$

Considering the electrostatic potential induced by the membrane at the full non-linear level, we allow only for a weakly charged polymer, so that its effect can be treated on the Debye-Hückel level. Assuming that we can superimpose the respective contributions to the electrostatic potentials as, returning to unscaled variables,

$$\psi_0(\mathbf{r}) = \psi_{0m}(\mathbf{r}) + \psi_{0p}(\mathbf{r}) \tag{2.112}$$

we end up with the variational equations

$$\nabla\varepsilon(\mathbf{r}) \cdot \nabla\psi_{0m}(\mathbf{r}) - \frac{2\varrho_b q e^2}{k_B T} e^{-V_i(\mathbf{r})} \sinh(q\psi_{0m}(\mathbf{r})) = -\frac{e^2}{k_B T}\sigma_m(\mathbf{r}), \tag{2.113}$$

$$\nabla\varepsilon(\mathbf{r}) \cdot \nabla v(\mathbf{r}, \mathbf{r}') - \frac{2\varrho_b q^2 e^2}{k_B T} e^{-V_i(\mathbf{r})} \cosh(q\psi_{0m}(\mathbf{r})) v(\mathbf{r}, \mathbf{r}') = -\frac{e^2}{k_B T}\delta(\mathbf{r} - \mathbf{r}'), \tag{2.114}$$

and

$$\left[\nabla\varepsilon(\mathbf{r}) \cdot \nabla - \frac{2\varrho_b q^2 e^2}{k_B T} e^{-V_i(\mathbf{r})} \cosh(q\psi_{0m}(\mathbf{r})) \right] \psi_{0p}(\mathbf{r}) = -\frac{e^2}{k_B T}\sigma_p(\mathbf{r}). \tag{2.115}$$

The comparison of the last two equations shows that one can interpret them in terms of an inverse kernel $v^{-1}(\mathbf{r}, \mathbf{r}')$ as

$$\int d\mathbf{r}' v^{-1}(\mathbf{r}, \mathbf{r}')\psi_{0p}(\mathbf{r}') = \sigma_p(\mathbf{r}). \tag{2.116}$$

If one now invokes the definition of the Green's function

$$\int d\mathbf{r}'' v^{-1}(\mathbf{r}, \mathbf{r}'')v(\mathbf{r}'', \mathbf{r}') = \delta(\mathbf{r} - \mathbf{r}'), \tag{2.117}$$

one can invert Eq. (2.116) to obtain

$$\psi_{0p}(\mathbf{r}) = \int d\mathbf{r}' v(\mathbf{r}, \mathbf{r}')\sigma_p(\mathbf{r}'). \tag{2.118}$$

Thus, to the level of our approximation we can write the grand potential of the polymer next to the membrane as the sum

$$\Omega_p = \Omega_{pm} + \Omega_{pp} \tag{2.119}$$

with

$$\Omega_{pm} = \int d\mathbf{r} \sigma_p(\mathbf{r}) \psi_{0m}(\mathbf{r}) \tag{2.120}$$

and

$$\Omega_{pp} = \frac{1}{2} \int d\mathbf{r} d\mathbf{r}' \sigma_p(\mathbf{r}) v(\mathbf{r}, \mathbf{r}') \sigma_p(\mathbf{r}'). \tag{2.121}$$

In order to compute these expressions, we need to solve Eqs. (2.113) and (2.114). For the configuration shown in Fig. 2.3, the mean-field potential in the half-space $z < 0$ for the charge density $\sigma_m(\mathbf{r}) = -\sigma_m \delta(z)$ is given by

$$\psi_{0m}(z) = -\frac{2}{q} \ln \left[\frac{1 + e^{\kappa_b(z-z_0)}}{1 - e^{\kappa_b(z-z_0)}} \right] \tag{2.122}$$

with $z_0 = -\ln[\gamma_c(s)]/\kappa_b$ with $\gamma_c(s) = \sqrt{s^2 + 1} - s$, where $s \equiv \kappa_b \ell_{GC}$ is the dimensionless ration of the Gouy-Chapman length and the Debye length.

For the calculation of the Green's function, making use of the planar geometry one can Fourier expand it in the form

$$v(\mathbf{r}, \mathbf{r}') = \int \frac{d^2\mathbf{k}}{4\pi^2} e^{i\mathbf{k} \cdot (\mathbf{r}_\parallel)} v(z, z'; k) \tag{2.123}$$

and introducing the permittivity profile

$$\varepsilon(z) = \varepsilon_w(-z) + \varepsilon_m \theta(z) \tag{2.124}$$

one obtains with the insertion of the result (2.122) the expression

$$\left[\partial_z \varepsilon_z \partial_s - \varepsilon_m \theta(z) - \varepsilon_s \theta(-z)[p^2 + 2\kappa_b \mathrm{csch}^2(\kappa_b(z - z_0))] \right] v(z, z'; k)$$
$$= -\frac{e^2}{k_B T} \delta(z - z'), \tag{2.125}$$

where csch denotes the hyperbolic cosecant and $p^2 = \sqrt{k^2 + \kappa_b^2}$. The full solution of this equation can be written as

$$v(z, z'; k) = c_1 h_-(z)\theta(-z)\theta(z' - z) + [c_2 h_-(z)$$
$$+ c_3(h_+(z)]\theta(-z)\theta(z - z') + c_4 e^{-kz}\theta(z), \tag{2.126}$$

with the increasing and decreasing homogeneous solution $h_\pm(z)$ given by

$$h_\pm = e^{\mp z}\left[1 \pm \frac{\kappa_b}{p}\coth(\kappa_b(z - z_0))\right].\tag{2.127}$$

The amplitudes $c_{1..4}$ in Eq. (2.126) can be determined from the boundary conditions. The first two require the continuity of v at $z = 0$ and $z = z'$, the third the continuity of the derivative $\varepsilon(z)\partial_z v$ at $z = 0$, and the final fourth condition is given by

$$\partial_z v(z = z'_+) - \partial_z v(z = z'_-) = -4\pi\ell_B.\tag{2.128}$$

The final result for $v(z, z'; k)$ can be summarized by

$$v(z, z'; k) = \frac{2\pi\ell_B p}{k^2}\left[h_+(z_>) + \Delta(k)h_-(z_>)\right]h_-(z_<),\tag{2.129}$$

whereby the index symbols denote $z_< = \min(z, z'), z_> = \max(z, z')$ and the function $\Delta(k)$ is given by

$$\Delta(k) = \frac{\kappa_b^2\text{csch}^2(\kappa_b z_0) + (p - \eta k)[p - \kappa_b\coth(\kappa_b z_0)]}{\kappa_b^2\text{csch}^2(\kappa_b z_0) + (p + \eta k)[p - \kappa_b\coth(\kappa_b z_0)]}.\tag{2.130}$$

Here, $\eta \equiv \varepsilon_m/\varepsilon_w$ characterizes the dielectric discontinuity at $z = 0$.

We now use these results to determine the adsorption of the polymer to the charged membrane. This requires some minor modifications to our expressions. First we place a polymer of length L at a distance z_a from the membrane, with its center of mass located at $y = 0$. Thus

$$\sigma_p(\mathbf{r}) = -\tau\theta(x + L/2)\theta(L/2 - x)\delta(y)\delta(z - z_a),\tag{2.131}$$

where τ is the *line charge density* of the polymer. We find Ω_{pm} to be given by

$$\Omega_{pm}(z_a) = \frac{2L\tau}{q}\ln\left[\frac{1 + e^{\kappa_b(z - z_0)}}{1 - e^{\kappa_b(z - z_0)}}\right].\tag{2.132}$$

Instead of Ω_{pp} we calculate a renormalized self-energy energy expression by calculating the quantity

$$\Delta\Omega_p = \Delta\Omega_{pp} + \Omega_{pm}\tag{2.133}$$

with

$$\Delta\Omega_{pp} = \frac{1}{2}\int d\mathbf{r}d\mathbf{r}'\sigma_p(\mathbf{r})[v(\mathbf{r}, \mathbf{r}') - v(\mathbf{r} - \mathbf{r}')]\sigma_p(\mathbf{r}')\tag{2.134}$$

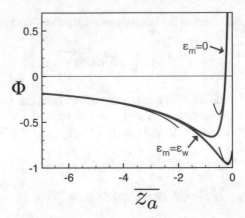

Fig. 2.4 The non-dimensionalized function $\Phi(\bar{z}_a)$ from Eq. (2.135) for two values of the membrane permittivity, shown in red and blue. The black curve describe analytic approximations to the full curve, which itself can be calculated analytically, see Ref. [5]. The main characteristics of the curve is the existence of a minimum at a finite value of $\bar{z}_a^*$, whose depth depends on the permittivity contrast between solvent and membrane. See the discussion in the text. © American Physical Society

where we subtract the Debye-Hückel Green's function. The explicit calculation of this expression is quite involved and will not be reproduced here; we refer to Ref. [5] for details. As the simplest case, the salt-free limit $\kappa_b \to 0$ can be considered. The final result in this case can be expressed as

$$\Delta\Omega_p(\bar{z}_a) = -\frac{2Q_p}{q} \ln(1 + \bar{z}_a) + \frac{\Xi_p}{2}\Phi(\bar{z}_a) \tag{2.135}$$

where $Q_p = L\tau$ and $\Xi_p = Q_p^2 \ell_B/\ell_{GC}$ assemble the characteristic lengths of the problem; the length z_a has been normalized with the Gouy-Chapman length, $\bar{z}_a \equiv z_a/\ell_{GC}$. The function Φ has been calculated analytically in Ref. [5], yielding a very complex expression, which we refrain from reproducing here. Figure 2.4 plots its behaviour. The main feature of the function is the existence of a minimum at a finite value of the distance from the membrane.

2.7 Summary

In this chapter we have seen how the Poisson-Boltzmann equation can be derived from a statistical physics approach, starting from a 'microscopic' description of the charged system. In this approach, the Poisson-Boltzmann equation arises as the saddle-point of the partition function, and is properly defined as a mean-field equation. We have then also seen how corrections to the Poisson-Boltzmann equation can be computed, and we looked at the lowest-order correction, the so-called one loop term. Practically, we employed this approach to the problem of ions near the water-air interface where a reduction of the surface tension is observed. This is a classic problem of physical chemistry, originally studied much earlier by Onsager and Samaras [1].

The following sections have taken the statistical mechanics of Poisson-Boltzmann theory in further directions. First we discussed an alternative approach to introduce fluctuation effects into PB-theory, a variational method. We then applied this method to the interaction of a stiff polymer with a rigid membrane. This first example has been a useful preparation for a more complex situation which we will encounter in the following chapter, when we will go beyond the assumption of a static dielectric constant for the solvent. Our next step in generalizing Poisson-Boltzmann theory thus consists in making the solvent properties 'explicit'.

2.8 Further Reading

The statistical physics of the Poisson-Boltzmann equation has been developed systematically in a series of papers by Netz and Orland [9–11]. Of most relevance is Ref. [11], as it describes the derivation of the full loop expansion around the saddle-point (the Poisson-Boltzmann equation itself).

For other approaches to calculate the fluctuation determinant, see the papers by Kirsten et al. [12–14].

One-loop corrections to the Poisson-Boltzmann equation have been derived by several authors; an example is the calculation of the static van der Waals forces by Netz [15]. Fluctuation effects in charge-regulated systems were discussed by Adžić and Podgornik [16]. One-loop corrections were also computed for extended Poisson-Boltzmann models, including the explicit solvent in a dipolar model discussed in the next chapter, see [17]. Finally, fluctuation effects in soap films have been discussed in a series of papers by Dean and Horgan [18–20]. In [20], the authors went to two-loop order.

Following the work by Netz and Orland [4], several authors have taken up the method and applied it to electrolyte systems. S. Buyukdagli and collaborators took it up in [21] and continued to publish papers on its applications, see [8, 22–24]. In the section on polymer adsorption we only scratched the surface of the topic: many further results can be found in Ref. [5]. A review on the topic is [25].

The variational method will receive considerably more attention in the next chapter. In the text we introduced the coupling parameter in passing. Its discussion goes beyond this text. We refer to the review by Naji et al. for a thorough discussion [26].

References

1. Onsager, L., Samaras, N.N.T.: The surface tension of Debye-Hückel electrolytes. J. Chem. Phys. **2**, 528–536 (1934)
2. Markovich, T., Andelman, D., Podgornik, R.: Surface tension of electrolyte solutions: a self-consistent theory. EPL **106**, 16002 (2014). https://doi.org/10.1209/0295-5075/106/16002
3. Markovich, T., Andelman, D., Podgornik, R.: Surface tension of electrolyte interfaces: ionic specificity within a field-theory approach. J. Chem. Phys. **142**, 044702 (2015)
4. Netz, R.R., Orland, H.: Variational charge renomalization in charged systems. Eur. Phys. J. E **11**, 301–311 (2003)
5. Buyukdagli, S., Blossey, R.: Correlation-induced DNA adsoprtion on like-charged membranes. Phys. Rev. E **94**, 042502 (2016)
6. Dunne, G.V.: Functional determinants in quantum field theory. J. Phys. A: Math. Theor. **41**, 304006 (2008)
7. Xu, Z., Maggs, A.C.: Solving fluctuation-enhanced Poisson-Boltzmann equations. J. Comp. Phys. **275**:310–322 (2014)
8. Buyukdagli, S., Achim, C.V., Ala-Nissila, T.: Electrostatic correlations in inhomogeneous charged fluids beyond loop expansion. J. Chem. Phys. **137**, 104902 (2012)
9. Netz, R.R., Orland, H.: Field theory for charged fluids and colloids. Europhys. Lett. **45**, 726–732 (1999)
10. Netz, R.R., Orland, H.: One and two-component hard-core plasmas. Eur. Phys. J. E **1**, 67–73 (2000)
11. Netz, R.R., Orland, H.: Beyond Poisson-Boltzmann: fluctuation effects and correlation functions. Eur. Phys. J. E **1**, 203–214 (2000)
12. Kirsten, K., McKane, A.J.: Functional determinants by contour integration methods. Ann. Phys. **308**, 502–527 (2003)
13. Kirsten, K., McKane, A.J.: Functional determinants for general Sturm-Liouville problems. J. Phys. A: Math. Gen. **37**, 4649–4670 (2004)
14. Kirsten, K., Loya, P.: Calculation of determinants using contour integrals. Am. J. Phys. **76**, 60–64 (2008)
15. Netz, R.R.: Static van der Waals interactions in electrolytes. Eur. Phys. J. E **5**, 189–205 (2001)
16. Adžić, N., Podgornik, R.: Field-theoretic description of charge regulation interaction. Eur. Phys. J. E **37**, 49 (2014)
17. Lévy, A., Andelman, D., Orland, H.: Dipolar Poisson-Boltzmann approach to ionic solutions: a mean-field and loop expansion analysis. J. Chem. Phys. **139**, 164909 (2013); Correction: **149**, 109901 (2018)
18. Dean, D.S., Horgan, R.: Electrostatic fluctuations in soap films. Phys. Rev. E **65**, 061603 (2002)
19. Dean, D.S., Horgan, R.: Weak nonlinear surface-charging effects in electrolytic films. Phys. Rev. E **68**, 051104 (2003)
20. Dean, D.S., Horgan, R.: Resummed two-loop calculation of the disjoining pressure of a symmetric electrolyte film. Phys. Rev. E **70**, 011101 (2004)
21. Buyukdagli, S., Manghi, M., Palmeri, J.: Variational approach for electrolyte solutions: from dielectric interfaces to charged nanopores. Phys. Rev. E **81**, 041601 (2010)
22. Buyukdagli, S., Achim, C.V., Ala-Nissila, T.: Ion size effects upon ionic exclusion from dielectric interfaces and slit nanopores. J. Stat. Mech. P05033 (2011)
23. Buyukdagli, S., Ala-Nissila, T.: Microscopic formulation of nonlocal electrostatics in polar liquids embedding polarizable ions. Phys. Rev. E **87**, 063201 (2013)
24. Buyukdagli, S., Blossey, R.: Dipolar correlations in structured solvents under nanoconfinement. J. Chem. Phys. **140**, 234903 (2014)
25. Buyukdagli, S., Blossey, R.: Beyond Poisson-Boltzmann: fluctuations and fluid structure in a self-consistent theory. J. Phys. Cond. Matter **28**, 343001 (2016)
26. Naji, A., Kanduc, M., Forsman, J., Podgornik, R.: Coulomb fluids-weak coupling, strong coupling, in between and beyond. J. Chem. Phys. **139**, 150901 (2013)

Chapter 3
Poisson-Boltzmann Theory with Solvent Structure

In this chapter we will further lift one key simplification made in the Poisson-Boltzmann equation which is the consideration of the solvent as a structureless continuum, or as an *implicit solvent*. The solvent is thus turned into an *explicit solvent* that is described by its intrinsic degrees of freedom.

There are essentially two ways how a step towards the introduction of explicit water structure can be achieved that have been discussed in the literature. The first approach is called *nonlocal electrostatics* and starts from the experimental fact that the dielectric behaviour of the solvent is not entirely captured by a macroscopic dielectric constant of $\varepsilon \approx 80$. The dielectric function or permittivity in general is a space and time- or frequency-dependent quantity, and, furthermore, also not a simple scalar: it needs to be considered as a *dielectric tensor* if space is not translationally invariant, which is the case in all systems we considered so far. This approach will be discussed in the first section.

The further sections then formulate a dipolar solvent model in a statistical mechanical theory for both the solvent molecules and the ions, in which to the lowest order the solvent molecules are treated as *point dipoles*. We then generalize the model to *finite size dipoles*. The dipole size thus enters as another length scale in the problem and indeed leads to a nonlocal version of the Poisson-Boltzmann equation. The resulting model is then first treated in the mean-field approximation and subsequently with the variational method.

At the end of the chapter, we will return to a simple phenomenological mean-field model that will prepare us for the Conclusions.

3.1 Nonlocal Electrostatics

For a general static dielectric function ε we have to consider a *nonlocal*, i.e. integral relationship for the dielectric displacement field

$$\mathbf{D}(\mathbf{r}) = \int d\mathbf{r}' \varepsilon(\mathbf{r}, \mathbf{r}') \mathbf{E}(\mathbf{r}') \tag{3.1}$$

in which the dielectric function $\varepsilon(\mathbf{r}, \mathbf{r}') = \varepsilon_{\alpha\beta}(\mathbf{r}, \mathbf{r}')$ is a tensor. We thus modify the left-hand side of our starting point, the Maxwell equation, and end up with a considerably more complex Poisson-Boltzmann equation than the standard equation, given by the *integro-differential equation*

$$-\nabla_{\mathbf{r}} \cdot \int d\mathbf{r}' \varepsilon(\mathbf{r}, \mathbf{r}') \nabla_{\mathbf{r}'} \phi(\mathbf{r}') = \varrho(\mathbf{r}), \tag{3.2}$$

$$\varepsilon(k) \equiv \left[1 - \frac{4\pi}{k_B T} S(k)\right]^{-1} \tag{3.3}$$

The necessity for this generalization of the Poisson-Boltzmann equation becomes clear when one considers the wave-vector dependence of the Fourier-transform of $\varepsilon(\mathbf{r}, \mathbf{r}')$ for bulk water, $\varepsilon(k)$. In the translationally invariant bulk ε is a scalar. Figure 3.1 displays experimental and simulation results of $\varepsilon(k)$ which in these cases is obtained via the relation between $\varepsilon(k)$ and the charge-density *structure factor* $S(k) = \langle \varrho_{-\mathbf{k}} \varrho_{\mathbf{k}} \rangle / k^2$. Indeed we see that the dielectric constant of the Poisson-Boltzmann equation holds in the *long-wavelength limit* $\lim_{k \to 0} \varepsilon(k)$. The complex behaviour of $\varepsilon(k)$ at finite wave-vector has been discussed in detail in the literature [1–3].

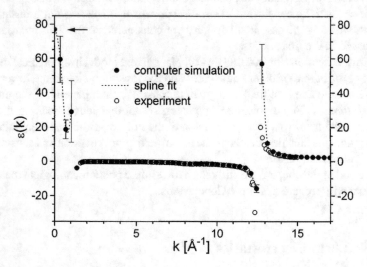

Fig. 3.1 Comparison between the simulated and the experimental longitudinal static dielectric function. From simulation data, $\varepsilon(k)$ was obtained from the static structure factor $S(k)$ that was calculated directly in Fourier space for a water model. The macroscopic value of the dielectric constant is $\varepsilon(k = 0) = 78$. Experimental data for $S(k)$ were derived from neutron diffraction data. The figure is taken from [1], where the details on the underlying computations can be found. © American Physical Society. Reprinted with permission

Fig. 3.2 A globular protein with its interior region Ω and the solvation region Σ. Figure taken from [7]. © American Physical Society

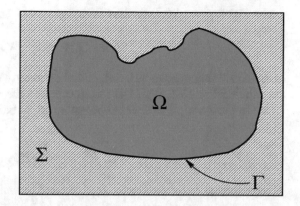

Solutions to the nonlinear Poisson Eq. (3.2) for the simple geometries we discussed in Chap. 1 have been discussed in [4–6] for several examples of approximate dielectric functions $\varepsilon(k)$. These computations are involved exercises in the theory of ordinary differential equations. Here we discuss the approach put forward by Hildebrandt et al. [7]. In this approach, one introduces a second potential, $\psi(\mathbf{r})$, which allows to compute the rotation-free part of the dielectric displacement field. Mathematically we have

$$\mathbf{D}(\mathbf{r}) = -\int d\mathbf{r}'\varepsilon(\mathbf{r},\mathbf{r}')\nabla_{\mathbf{r}'}\phi(\mathbf{r}) \equiv -\nabla\psi(\mathbf{r}) + \nabla\times\boldsymbol{\xi}(\mathbf{r}). \qquad (3.4)$$

In [7], the situation of globular proteins has been addressed, in which the rotational part can, to a good approximation, be ignored. One then has the to solve the problem

$$\Delta\psi(\mathbf{r}) = -\varrho(\mathbf{r}), \quad \mathbf{r}\in\Omega \qquad (3.5)$$

where Ω represents the approximately spherical protein interior, in which the partial charges of the protein are localized, see the illustration in Fig. 3.2.

In the outer region Σ, ψ fulfills the Poisson equation to solve the problem

$$\Delta\psi(\mathbf{r}) = 0, \quad \mathbf{r}\in\Sigma. \qquad (3.6)$$

The relation between $\phi(\mathbf{r})$ and $\psi(\mathbf{r})$ is provided by Eq. (3.4); both fields are furthermore coupled by the boundary conditions

$$\nabla\psi_{\Sigma}|_n = \nabla\psi_{\Omega}|_n \text{ and } \nabla\phi_{\Sigma}|_t = \nabla\phi_{\Omega}|_t \qquad (3.7)$$

for the normal and tangential components. Finally, in region Ω, we have $\varepsilon_{\Omega}\phi_{\Omega} = \psi_{\Omega}$.

The key to further progress now is an assumption on $\varepsilon(\mathbf{r},\mathbf{r}')$. If we make the ansatz

$$\varepsilon(\mathbf{r},\mathbf{r}') = \varepsilon_{\ell}\delta(\mathbf{r}-\mathbf{r}') + \tilde{\varepsilon}G(\mathbf{r},\mathbf{r}') \qquad (3.8)$$

where $\tilde{\varepsilon} = (\varepsilon_\Sigma - \varepsilon_\ell)/\lambda^2$ and the Green function G fulfills

$$\mathcal{L}G(\mathbf{r}, \mathbf{r}') = -\delta(\mathbf{r} - \mathbf{r}') \tag{3.9}$$

in which $\mathcal{L}$ is assumed to be a differential operator with constant coefficients. With the introduction of $\mathcal{L}$, Eq. (3.4) transforms into the ordinary differential equation (since in this case, all differential operators are to be assumed to be taken in radial coordinates)

$$(\varepsilon_\ell \mathcal{L} - \tilde{\varepsilon})\phi_\Sigma = -\mathcal{L}\psi_\Sigma. \tag{3.10}$$

A standard choice for the Green function is a Yukawa-type kernel in real space

$$G(\mathbf{r} - \mathbf{r}') = \frac{1}{4\pi} \frac{e^{-|\mathbf{r}-\mathbf{r}'|/\lambda}}{|\mathbf{r} - \mathbf{r}'|} \tag{3.11}$$

so that

$$\mathcal{L} \equiv \Delta - \frac{1}{\lambda^2}. \tag{3.12}$$

Thus, one finally arrives at the equation

$$(\varepsilon_\ell \lambda^2 \Delta - \varepsilon_\Sigma)\phi_\Sigma(\mathbf{r}) = \psi_\Sigma(\mathbf{r}). \tag{3.13}$$

This equation resembles formally a Debye-Hückel equation, the linearized version of the Poisson-Boltzmann equation, in which now the Debye-Hückel length is played by the combination of dielectric constant and the length λ which we interpret loosely as a correlation length for water orientations. The potential ψ acquires the role of the density of mobile charges. The field ψ can thus be interpreted as a *density of polarization charges* in the bulk, whose gradient generates the rotation-free part of the displacement field.

This theory has been applied to compute the electrostatic potential of proteins [8], which requires a substantial effort in the discretization of the equation for a very inhomogenous boundary problem. We will not further pursue this line of research here, and rather continue with a different approach based on statistical physics.

Nonlocal electrostatics as we have written it down before is *phenomenological*: we represent the dielectric properties by some function which approximates solvent properties; this comes along with the introduction of a number of parameters, like the additional length λ introduced before. In fact, the complex structure of the dielectric function $\varepsilon(k)$ only comes about because there are additional length scales in the problem. The long-wavelength limit $k \to 0$ washes all these microscopic lengths out. But ultimately the solvent *is* a grainy object: made from molecules (water), that, moreover show correlations on larger scales than their size, which necessarily lead to the consideration of further length scales.

In the next sections, we build up such models from a microscopic point of view.

3.2 The Dipolar Poisson-Boltzmann Equation: A Point-Dipole Theory

In this section of we will introduce the *dipolar Poisson-Boltzmann equation* (in short, DPB-equation) that considers a system of N_d mobile point dipoles with dipole moment $\mathbf{p}$ and in general I ion species (practically, we will always only consider $I = 2$). They are considered in a continuum dielectric background with dielectric constant ϵ that, while larger than than the vacuum permittivity ε_0, is not equal to the dielectric constant of water, because of the contribution of the solvent molecular dipoles—as we will see. The total charge density of this system is given by, following [9]

$$\varrho(\mathbf{r}) = -\sum_{i=1}^{N_d} \mathbf{p}_i \cdot \nabla \delta(\mathbf{r} - \mathbf{r}_i) + \sum_{j=1}^{I} \sum_{i=1}^{N_j} q_j e \delta(\mathbf{r} - \mathbf{R}_i^{(j)}) + \varrho_f(\mathbf{r}). \tag{3.14}$$

Here, $q_j e$ is the charge of the j-th ion. The terms refer to the solvent dipoles, the ions and the fixed charge distribution, respectively.

We have seen in the previous chapter how to write down the partition function for such a system. It is given by

$$Z = \int \mathcal{D}[\mathbf{r}, \mathbf{p}, \mathbf{R}] e^{-\beta/2 \int d^3\mathbf{r} d^3\mathbf{r}' \varrho(\mathbf{r}) v_c(\mathbf{r} - \mathbf{r}') \varrho(\mathbf{r}')}, \tag{3.15}$$

where v_c is the Coulomb potential and the integration measure is defined by

$$\mathcal{D}[\mathbf{r}, \mathbf{p}, \mathbf{R}] \equiv \frac{1}{N_d! \prod_{j=1}^{I} N_j!} \prod_{i=1}^{N_j} d^3\mathbf{r}_i d^3\mathbf{p}_i \prod_{j=1}^{I} \prod_{i=1}^{N_j} d^3\mathbf{R}_i^{(j)}. \tag{3.16}$$

Making use of the Hubbard-Stratonovich transform we reformulate this involved expression as a path integral over the electrostatic potential, as before. We find $Z = \int \mathcal{D}\phi e^{-\beta H}$ with

$$\beta H = \left(\frac{\beta \epsilon}{2} \int d^3\mathbf{r} (\nabla \phi(\mathbf{r}))^2 - \lambda_d \int d^3\mathbf{r} d^3\mathbf{p} \, e^{-i\beta \mathbf{p} \cdot \nabla \phi} \right.$$
$$\left. - \sum_{i=1}^{I} \lambda_i \int d^3\mathbf{r} \, e^{-i\beta q_i e\phi} + i\beta \int d^3\mathbf{r} \phi(\mathbf{r}) \varrho_f(\mathbf{r}) \right). \tag{3.17}$$

The fugacities λ_k for $k = i, d$ derive from the relation $N_k = \lambda_k (\partial/\partial \lambda_k) \log Z$.

The expression can be simplified by assuming a fixed magnitude for the point dipoles, $\mid \mathbf{p} \mid = p_0$. The integrals over the dipole momenta $\{\mathbf{p}\}$ can then be computed, leading to a term

$$\sim \lambda_d \int d^3 \mathbf{r} \frac{\sin(\beta p_0 |\nabla \phi|)}{\beta p_0 |\nabla \phi|}.$$

The DPB-equation is, as before the standard Poisson-Boltzmann equation, obtained by the computation of the saddle-point of the partition function Z. Introducing the physical potential by setting $\phi \to i\phi$ we obtain the mean-field saddle-point equation

$$-\epsilon \nabla^2 \phi = \sum_i \lambda_i (q_i e) e^{-\beta q_i e \phi} + \lambda_d p_0 \nabla \cdot \left(\frac{\nabla \phi}{|\nabla \phi|} \mathcal{G}(\beta p_0 |\nabla \phi|) \right) + \varrho_f(\mathbf{r}). \quad (3.18)$$

The function $\mathcal{G}(u)$ is given by

$$\mathcal{G}(u) = \frac{\cosh u}{u} - \frac{\sinh u}{u^2}. \quad (3.19)$$

It is related to the *Langevin function* $\mathcal{L}(u)$ via

$$\mathcal{G}(u) = \frac{\sinh u}{u} \mathcal{L}(u) \quad (3.20)$$

where

$$\mathcal{L}(u) = \coth u - \frac{1}{u}. \quad (3.21)$$

We can now essentially restart the machinery and discuss the solutions of the DPB-equation in simple geometries, as we discussed in Chap. 1 for the standard PB-equation. In one dimension, the partial differential equation simplifies to

$$-\epsilon \phi''(z) = -2(c_i e) \sinh(\beta e \phi) + c_d p_0 \frac{d}{dz} \mathcal{G}(\beta p_0 \phi'(z)) + \varrho_f(z) \quad (3.22)$$

where we have replaced fugacities by concentrations, which is correct at mean-field level. Ignoring the ions, this equation can be rewritten in terms of the electric field $E(z) = d\phi/dz$ as

$$-\frac{d}{dz} \left(\epsilon E(z) + c_d p_0 \mathcal{G}(\beta p_0 E(z)) \right) = \varrho_f(z). \quad (3.23)$$

We can therefore introduce an *effective field-dependent dielectric function* by setting

$$\epsilon^{eff} \equiv \epsilon + \frac{c_d p_0}{E} \mathcal{G}(\beta p_0 E). \quad (3.24)$$

Putting the ions back and expanding the function $\mathcal{G}$ for weak fields, one finds back the standard PB-equation

$$\epsilon^{eff} \phi''(z) \approx 2(c_i e) \sinh(\beta e \phi(z)) \quad (3.25)$$

with an effective dielectric constant

$$\epsilon^{eff} = \epsilon + \frac{\beta c_d p_0^2}{3}. \tag{3.26}$$

This equation explains why we chose ϵ as being larger than the vacuum permittivity—evident—but smaller than water permittivity: the explicit dipoles now contribute to the overall dielectric constant. The physical value of the dipole moment for water is $p_0 = 1.85$ D. In order to reach the values of water, we however must crank up the dipole moment by a factor of about 2.5; for $p_0 = 4.86$ D we obtain $\epsilon_r^{eff} \approx 80$ for a water molecular concentration of $c_d = 55$ M.

The DPB-theory has one further important modification, which is needed when the dipolar effects are strong and the resulting ion and dipole densities can become unphysically large. This can be avoided in the theory by putting the constituent ions and dipoles on a lattice of spacing ℓ on the order of their molecular size. The built-in incompressibility by the lattice constraint then leads to the expression at mean-field level

$$-\beta F = \frac{\beta \epsilon}{2} \int d^3\mathbf{r} (\nabla \phi(\mathbf{r}))^2$$
$$+ \frac{1}{\ell^3} \int d^3\mathbf{r} \log \left(c_d \frac{\sinh(\beta p_0 |\nabla \phi|)}{\beta p_0 |\nabla \phi|} + 2c_s \cosh(\beta e \phi) \right), \tag{3.27}$$

where the lattice constraint imposes the condition $c_d + 2c_s = \ell^{-3}$.

3.3 Finite-Size Dipoles

Having introduced point dipoles into Poisson-Boltzmann theory in Chap. 2, the logical next step is to turn them into finite-size dipoles. This provides a physically unambiguous introduction of an additional length scale into Poisson-Boltzmann theory, and will allow us to 'marry' the nonlinear DPB-theory to nonlocal electrostatics, and not just in a phenomenological way, but in a 'clean' statistical mechanics approach.

We thus now introduce a model composed of a symmetric electrolyte of ions immersed in a polar solvent composed of dipole molecules of finite size that carry two elementary charges of opposite signs $+Q$ and $-Q$ separated by a distance a, as shown in Fig. 3.3.

We write down the partition function for this model as $Z = \int \mathcal{D}\phi \, e^{-H[\phi]}$ with the Hamiltonian functional $H[\phi]$, originally introduced in Ref. [10], as

$$H[\phi] = \int d\mathbf{r} \left[\frac{[\nabla \phi(\mathbf{r})]^2}{8\pi \ell_B} - i\sigma(\mathbf{r})\phi(\mathbf{r}) \right]$$
$$- \Lambda_s \int \frac{d\mathbf{r} d\Omega}{4\pi} e^{i Q[\phi(\mathbf{r}) - \phi(\mathbf{r+a})]} - \int d\mathbf{r} \left\{ \Lambda_+ e^{iq\phi(\mathbf{r})} + \Lambda_- e^{-iq\phi(\mathbf{r})} \right\}, \tag{3.28}$$

Fig. 3.3 The geometry of
the solvent molecules with
size $a = 1$ Å; the charges of
valency $Q = 1$ are placed at
the ends of a linear rod

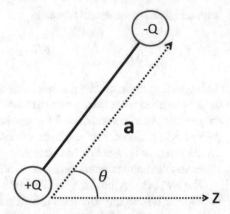

in which the ionic and solvent fugacities Λ_i with $i = \pm$ and Λ_s. The second term in this expression describes the explicitly introduced solvent molecules.

In Eq. (3.28) temperature is included in the Bjerrum length in air which is given by $\ell_B = e^2/(4\pi\varepsilon_{air}k_B T)$ and $\sigma(\mathbf{r})$ is the fixed surface charge distribution. If we expand the exponential of the dipolar term in the limit $\mathbf{a} \to 0$, we recover the dipolar Poisson-Boltzmann (DPB) model of point dipoles.

In the same way as we discussed the standard Poisson-Boltzmann theory in Chap. 2, we proceed first with the discussion of the mean-field solution. Passing from the complex to the real electrostatic potential by setting $\phi(\mathbf{r}) \to i\phi(\mathbf{r})$, the mean-field saddle point equation $\delta H[\phi]/\delta\phi(\mathbf{r}) = 0$ takes the form of a *nonlocal Poisson-Boltzmann (NLPB) equation*.

We will consider this model here in a system with a single plane considered permeable to the molecules. Its charge distribution is given by

$$\sigma(\mathbf{r}) = -\sigma_s\delta(z).\tag{3.29}$$

In this one-dimensional geometry, the NLPB-equation is given by the nonlocal expression

$$\Delta\phi(z) + 4\pi\ell_B\sigma(z) - 8\pi\ell_B\rho_i^b q \sinh\left[q\phi(z)\right]$$

$$+8\pi\ell_B Q\rho_s^b \int_{-a}^{a} \frac{da_z}{2a} \sinh\left[Q\phi(z+a_z) - Q\phi(z)\right] = 0,\tag{3.30}$$

where we have employed the mean-field relations $\Lambda_i = \rho_i^b$ and $\Lambda_s = \rho_s^b$ for the ion and dipole bulk densities, respectively.

In the regime of weak surface charges, where $\phi(z) \ll 1$, a linearization of the NLPB-equation (3.30) is allowed. It results in the expression

$$\Delta\phi_0(z) - \kappa_i^2 \phi(z) + \kappa_s^2 \int\limits_{-a}^{a} \frac{da_z}{2a} \left[\phi_0(z + a_z) - \phi_0(z)\right] = -4\pi\ell_B\sigma(z), \qquad (3.31)$$

where the index '0' at the electrostatic potential has been introduced to indicate that nonlinearities have been dropped. Furthermore, we introduce the ion and solvent screening parameters in air as $\kappa_i^2 = 8\pi\ell_B\rho_i^b q_i^2$ and $\kappa_s^2 = 8\pi\ell_B\rho_s^b Q^2$, respectively.

Since we do not consider the system as strictly bounded by a wall, we can solve Eq. (3.31) in Fourier space. In the parameter range $0 \leq \kappa_i z \ll 1$ the electric field $E(z) = \partial_z \phi(z)$ fulfills

$$E(z) = \frac{2\pi\ell_B\sigma_s}{\varepsilon_{eff}(z)}, \qquad (3.32)$$

in which the *effective dielectric permittivity*

$$\varepsilon_{eff}(z) = \frac{\pi}{2} \left[\int\limits_{0}^{\infty} \frac{dk}{k} \frac{\sin(kz)}{\varepsilon(k)} \right]^{-1} \qquad (3.33)$$

has been introduced with the Fourier-transformed permittivity

$$\varepsilon(k) = 1 + 4\pi\ell_B\chi_0(k) \qquad (3.34)$$

wherein the *susceptibility function* $\chi_0(k)$ is given by

$$\chi_0(k) = \frac{\kappa_s^2}{4\pi\ell_B k^2} \left[1 - \frac{\sin(ka)}{ka} \right]. \qquad (3.35)$$

Equation (3.33) clarifies the concept of a *distance-dependent effective permittivity* which is often used phenomenologically in the literature, see, e.g., [11]. To the leading order in the solvent density $O\left((\kappa_s a)^2\right)$, the effective permittivity follows from Eq. (3.33) in the simple form

$$\varepsilon_{eff}(z) = 1 + \frac{(\kappa_s a)^2}{6} \left\{ 1 - \left(1 - \frac{z}{a}\right)^3 \theta(a - z) \right\}. \qquad (3.36)$$

Thus, for dilute solvents and in the *linear response regime*, the effect of the finite solvent molecular size is an increase of the effective permittivity from the air permittivity at the interface, where the polarization field vanishes, to the bulk permittivity $\varepsilon_w = 1 + (\kappa_s a)^2 / 6$.

We can now take a look at the nonlinear correction to this result that we write as, motivated by Eq. (3.36),

$$\varepsilon(k) = 1 + 4\pi\ell_B \left[\chi_0(k) + \delta\chi(k)\right]. \qquad (3.37)$$

Thus we first have to derive the nonlinear susceptibility function, which we will obtain from a perturbative approach. To this end, we first insert into the non-linear mean-field free energy (3.28) the Green-function based *ansatz*

$$\phi(z) = \int\limits_{-\infty}^{+\infty} dz' G(z - z') \sigma(z') \tag{3.38}$$

and expand the equation to second order in the surface charge. We then postulate for the Green function a formal expansion

$$G(z) = G_0(z) + \lambda G_1(z), \tag{3.39}$$

where the perturbative parameter λ is introduced as a book-keeping parameter to compute the correction to the linear response solution; it will later be set to one. The zeroth-order Green function[1]

$$G_0(z) = \frac{1}{2\pi} \int dk \, G_0(k) \tag{3.40}$$

solves the linear NLPB-equation (3.31) in the form

$$\phi_0(z) = \int\limits_{-\infty}^{+\infty} dz' G_0(z - z') \sigma(z') \tag{3.41}$$

with

$$G_0^{-1}(k) = \frac{\kappa_i^2 + k^2}{4\pi \ell_B} + k^2 \chi_0(k). \tag{3.42}$$

The correction $G_1(k)$ to the Green's function $G_0(k)$ satisfies the differential equation

$$\partial_z^2 G_1(z) - \kappa_i^2 G_1(z) - \kappa_s^2 \int\limits_{-a}^{a} \frac{da_z}{2a} \left[G_1(z) - G_1(z + a_z) \right]$$

$$= \frac{\sigma_s^2}{6} \left\{ q^2 \kappa_i^2 G_0^3(z) + \kappa_s^2 Q^2 \int\limits_{-a}^{+a} \frac{da_z}{2a} \left[G_0(z) - G_0(z + a_z) \right]^3 \right\}. \tag{3.43}$$

which is linear in the nonlinear correction G_1 and nonlinear in G_0. Solving Eq. (3.43) again in Fourier space, the nonlinear correction is obtained as

---

[1] We use the same function symbol for the Fourier-transform and its inverse, the meaning should be clear from its argument.

$$G_1(k) = -\frac{\sigma_s^2}{24\pi \ell_B} G_0(k) \left\{ q^2 \kappa_i^2 F(k) + Q^2 \kappa_s^2 T(k) \right\}, \tag{3.44}$$

where we defined the functions

$$F(k) = \int\limits_{-\infty}^{+\infty} \frac{dk_1 dk_2}{4\pi^2} G_0(k_1) G_0(k_2) G_0(k - k_1 - k_2) \tag{3.45}$$

$$T(k) = \int\limits_{-\infty}^{+\infty} \frac{dk_1 dk_2}{4\pi^2} G_0(k_1) G_0(k_2) G_0(k - k_1 - k_2)$$
$$\times R(k_1, k_2, k - k_1 - k_2), \tag{3.46}$$

with the structure factor R being given by the involved expression

$$R(k_1, k_2, k_3) = 1 - \sum_{i=1}^{3} \frac{\sin(k_i a)}{k_i a} + \frac{\sin\left[(k_1 + k_2)a\right]}{(k_1 + k_2)a}$$
$$+ \frac{\sin\left[(k_1 + k_3)a\right]}{(k_1 + k_3)a} + \frac{\sin\left[(k_2 + k_3)a\right]}{(k_2 + k_3)a}$$
$$- \frac{\sin\left[(k_1 + k_2 + k_3)a\right]}{(k_1 + k_2 + k_3)a}. \tag{3.47}$$

At the perturbative order $O(\lambda)$ the Fourier transform of the kernel associated with the Green's function (3.39) can be written as

$$G^{-1}(k) = G_0^{-1}(k) - \lambda G_0^{-2}(k) G_1(k). \tag{3.48}$$

Taking into account Eq. (3.42) and the nonlinear contribution in Eq. (3.44), one finds that the Fourier-transformed kernel can be expressed in the form

$$G^{-1}(k) = \frac{\kappa_i^2 + k^2 + \lambda \delta \kappa_i^2(k)}{4\pi \ell_B} + k^2 \left[\chi_0(k) + \lambda \delta \chi(k) \right], \tag{3.49}$$

which is worth comparing with Eq. (3.42). The newly appearing corrections to the screening and the susceptibility functions because of the nonlinearities are given by the expressions

$$\delta \kappa_i^2(k) = \frac{q^2 \sigma_s^2 \kappa_i^2}{6} \frac{F(k)}{G_0(k)} \tag{3.50}$$

Fig. 3.4 **a** The surface
charge dependence of the
dielectric permittivity
function of Eq. (3.37) and **b**
the linear (main plot) and
non-linear response part
(inset) of the susceptibility
function in Fourier space.
Bottom: Dielectric
permittivity profile in real
space. Figure taken from
[10]. © IOP Publishing.
Reproduced with permission.
All rights reserved

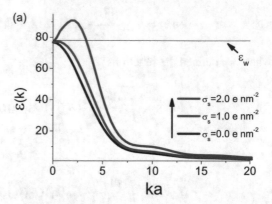

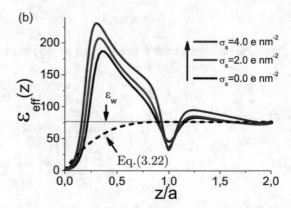

$$\delta\chi(k) = \frac{Q^2\sigma_s^2\kappa_s^2}{24\pi\ell_B}\frac{T(k)}{k^2G_0(k)}. \tag{3.51}$$

The relation (3.37) together with the new result Eq. (3.51) shows that, to leading
order, the dielectric permittivity has a quadratic dependence on the surface charge.

We can now visualize the dielectric functions $\varepsilon(k)$ that we obtained from this
nonlocal electrostatics theory within the mean-field approach. The model parameters are chosen as $a = 1$ Å, $Q = 1$ for the solvent molecules, the water density as
$\rho_s^b = 55$ M, and we ignore the ions by setting $\rho_i^b = 0$ M. This results in a value for
the bulk permittivity $\varepsilon_w = 77$, given by the horizontal lines in the graphs of Fig. 3.4
which shows the dielectric permittivity profiles in both Fourier- and real space for
finite surface charges. The latter was obtained via Eq. (3.32) from the numerical
solution of the non-linear NLBP Eq. (3.30) in the limit of vanishing salt density
$\rho_{ib} \to 0$; the details of the numerical relaxation algorithm can be found in the original reference [10].

First of all, it is seen that the modification of the permittivity by the surface charge becomes significant in the regime $\sigma_s \geq 1$ e nm^{-2}, the boundary value corresponding, e.g., to the characteristic charge of a DNA molecule. It turns out to, in fact, be located in a parameter regime which goes beyond the mean-field Poisson-Boltzmann regime—the discussion of this issue goes beyond what is presented here. In particular it means that the non-linear response behavior only comes into play in a surface charge regime in which correlation effects beyond mean-field aris—we will take this as a motivation to go beyond mean field in the next section.

Further, one notes that the permittivity increases with the surface charge, i.e., a dielectric saturation effect is not observed in our model. Finally, one notices that the surface charge makes no contribution to the permittivity function in the infrared (meaning the limit $k \to 0$) and the ultraviolet regime (meaning $k \to \infty$) which correspond to the bulk region and the close vicinity of the interface, respectively.

In the plot of the dependence of the effective dielectric permittivity $\varepsilon_{eff}(z)$ on the spatial distance from the charged interface one should first look at the dashed and solid black curves which corresponding to the vanishing surface charge limit $\sigma_s \to 0$. They correspond to the expressions (3.36) and (3.33), respectively. Non-local effects associated with the departure from the dilute solvent regime thus are seen to translate into strong variations of the local permittivity function around bulk permittivity. In qualitative agreement with the Fourier-transformed permittivity profiles in Fig. 3.4 the increase of the surface charge resulting in a deviation from the linear response regime is seen to increase the amplitude of the dielectric permittivity in real space. The overall shape of the effective permittivity profile is, however, not modified by the surface charge.

3.4 Fluctuation Effects in the Nonlocal Dipolar Poisson-Boltzmann Theory

In this section we will go beyond the mean-field solution to the nonlocal Poisson-Boltzmann theory presented in the previous section. This time, we will not employ the loop expansion, but instead make use of the *variational method* that in the context of the statistical mechanics of Poisson-Boltzmann theory was discussed in Chap. 2. Here we follow closely the variational approach as it was further developed and applied by Buyukdagli et al. [12–14].

We therefore apply the variational method to the nonlocal Hamiltonian of the dipolar solvent. The solvent charge structure and the composition of the charged fluid in the nanoslit are schematically drawn in Fig. 3.5. In principle we can allow the liquid to contain an arbitrary number of ionic species $i = 1, \ldots, p$ with each species having the valency q_i; we will stick to two. The Hamiltonian functional is given by the expression

Fig. 3.5 Slit geometry with
a surface charge $-\sigma_s < 0$
confining the solvent
molecules (blue dipoles),
anions (green circles) and
cations (red circles)

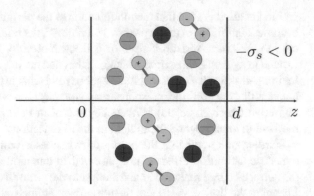

$$H[\phi] = \frac{k_B T}{2e^2} \int d\mathbf{r}\, \varepsilon_0(\mathbf{r})\, [\nabla\phi(\mathbf{r})]^2 - i \int d\mathbf{r}\sigma(\mathbf{r})\phi(\mathbf{r})$$
$$- \Lambda_s \int \frac{d\mathbf{r}d\Omega}{4\pi} e^{E_s - W_s(\mathbf{r},\mathbf{a})} e^{iQ[\phi(\mathbf{r})-\phi(\mathbf{r}+\mathbf{a})]}$$
$$- \sum_i \Lambda_i \int d\mathbf{r} e^{E_i - W_i(\mathbf{r})} e^{iq_i\phi(\mathbf{r})}, \tag{3.52}$$

where the function $\varepsilon_0(\mathbf{r})$ accounts for the dielectric permittivity difference between
vacuum and a membrane with permittivities ε_0 and ε_m, respectively. The dielectric
permittivities will be given in units of the vacuum permittivity, which is equivalent to
setting $\varepsilon_0 = 1$. However, for the sake of generality, the coefficient ε_0 will nevertheless
be kept in the equations.

In Eq. (3.52), the third and fourth terms correspond to the number density of
mobile ions and solvent molecules, respectively, and, as before, are modified with
respect to the previous expression by the introduction of self-energies of ions and
polar molecules in vacuum that are given by

$$E_i = \frac{q_i^2}{2} v_c(0), \quad E_s = Q^2 v_c(0). \tag{3.53}$$

In Eq. (3.53), the bulk Coulomb potential in vacuum is given by $v_c(r) = \ell_B/r$ with
$\ell_B = e^2/(4\pi\varepsilon_0 k_B T)$ is the Bjerrum length in vacuum. The nonlocal expression in
the exponential of the third term of Eq. (3.52) is the generalization of a term

$$\sim \int \frac{d\mathbf{r}d\Omega}{4\pi} e^{E_s - W_s(\mathbf{r},\mathbf{a})} e^{i\mathbf{p}\cdot\nabla\phi} \tag{3.54}$$

with the dipole moment $\mathbf{p}$ in the local case, see the DPB-theory in Sect. 3.2 of this
chapter. In taking the point-dipole limit of Eq. (3.52), which consists in Taylor-
expanding the dipolar term to quadratic order in the solvent molecular size a, one
obtains the Hamiltonian of the point-dipole liquid introduced in Ref. [9, 15]. Starting

from the *point-dipole model*, in Ref. [16] a variational (there called 'extended') dipolar Poisson-Boltzmann (EDPB) model has been developed, that incorporates interfacial correlation effects on the solvent dielectric response. e will include this model in our comparison with some results we obtain from the nonlocal variational model.

The Hamiltonian contains general wall potentials $W_i(\mathbf{r})$ and $W_s(\mathbf{r}, \mathbf{a})$ that for ions and solvent molecules account for the presence of the impenetrable boundaries in the system—in contrast to our previous setup in Sect. 3.3. Evaluating the field-theoretic averages as described for the variational method in Chap. 2 we obtain the expression of the variational *grand potential* in the form[2]

$$\Omega_v = -\frac{1}{2}\text{Tr}\ln[v_0] + \int d\mathbf{r}\sigma(\mathbf{r})\phi_0(\mathbf{r})$$
$$+ \frac{k_B T}{2e^2}\int d\mathbf{r}\left[\varepsilon_0(\mathbf{r})\nabla_\mathbf{r}\cdot\nabla_{\mathbf{r}'}\,v_0(\mathbf{r},\mathbf{r}')\big|_{\mathbf{r}'\to\mathbf{r}} - \varepsilon_0(\mathbf{r})\left[\nabla\phi_0(\mathbf{r})\right]^2\right]$$
$$- \sum_i \Lambda_i \int d\mathbf{r}\, e^{E_i - W_i(\mathbf{r})}e^{-q_i\phi_0(\mathbf{r})}e^{-\frac{q_i^2}{2}v_0(\mathbf{r},\mathbf{r})}$$
$$- \Lambda_s \int d\mathbf{r}\frac{d\Omega}{4\pi}e^{E_s - W_s(\mathbf{r},\mathbf{a})}e^{-Q[\phi_0(\mathbf{r})-\phi_0(\mathbf{r}+\mathbf{a})]}\,e^{-\frac{Q^2}{2}v_d(\mathbf{r},\mathbf{a})}, \tag{3.55}$$

where the first term resulting from the quadratic fluctuations, the *van der Waals term*, is given by

$$\Omega_0 = -\ln\int\mathcal{D}\phi\,e^{-H_0[\phi]} = -\frac{1}{2}\text{Tr}\ln[v_0]. \tag{3.56}$$

The forth and fifth terms on the right-hand side of the grand potential (3.55) correspond, respectively, to the average densities of ions and dipoles. In the fifth term we also introduced the *dipolar self-energy* defined as

$$v_d(\mathbf{r}, \mathbf{a}) = v_0(\mathbf{r}, \mathbf{r}) + v_0(\mathbf{r} + \mathbf{a}, \mathbf{r} + \mathbf{a}) - v_0(\mathbf{r}, \mathbf{r} + \mathbf{a}) - v_0(\mathbf{r} + \mathbf{a}, \mathbf{r}). \tag{3.57}$$

The number density of the ions is determined from the grand potential according to

$$\rho_i(\mathbf{r}) = \frac{\delta\Omega_v}{\delta W_i(\mathbf{r})} \tag{3.58}$$

which yields the expression

$$\rho_i(\mathbf{r}) = \Lambda_i e^{E_i - W_i(\mathbf{r})}e^{-q_i\phi_0(\mathbf{r})}e^{-\frac{q_i^2}{2}v_0(\mathbf{r},\mathbf{r})}. \tag{3.59}$$

---

[2] Note that we have changed the symbol for the Green function and covariance for which we now use the symbol v, in accord with the use of symbols in Ref. [14].

From the bulk limit of Eq. (3.59), the relation between the ion fugacity and the reservoir concentration is obtained as

$$\rho_{ib} = \Lambda_i \exp\left(E_i - \frac{q_i^2}{2} v_0^b(0) \right), \tag{3.60}$$

where $v_0^b(0)$ is the ion self-energy in a bulk solvent, also called the equal-point *electrostatic propagator* in the absence of boundaries. In order to obtain the number density of the two elementary charges located at the ends of the solvent molecule, we split the wall potential into two parts, $W_s(\mathbf{r}, \mathbf{a}) = W_+(\mathbf{r}) + W_-(\mathbf{r} + \mathbf{a})$, where the functions $W_+(\mathbf{r})$ and $W_-(\mathbf{r} + \mathbf{a})$ respectively are the steric potentials experienced by the negative and positive charges on the solvent molecule [the origin of the molecule located at $\mathbf{r}$ corresponds to the positive charge, see Fig. (3.5)]. By taking the functional derivatives of the grand potential (3.55) with respect to the potential $W_\pm(\mathbf{r}, \mathbf{a})$, the number densities for the solvent follow in the form

$$\rho_{s\pm}(\mathbf{r}) = \int \frac{d\Omega}{4\pi} f_{s\pm}(\mathbf{r}, \mathbf{a}), \tag{3.61}$$

with the solvent densities at a fixed orientation Ω defined as

$$\begin{aligned}
f_{s+}(\mathbf{r}, \mathbf{a}) &= \Lambda_s e^{E_s - W_s(\mathbf{r}, \mathbf{a})} e^{-Q[\phi_0(\mathbf{r}) - \phi_0(\mathbf{r}+\mathbf{a})]} e^{-\frac{Q^2}{2} v_d(\mathbf{r}, \mathbf{a})} \\
f_{s-}(\mathbf{r}, \mathbf{a}) &= \Lambda_s e^{E_s - W_s(\mathbf{r}-\mathbf{a}, \mathbf{a})} e^{-Q[\phi_0(\mathbf{r}-\mathbf{a}) - \phi_0(\mathbf{r})]} e^{-\frac{Q^2}{2} v_d(\mathbf{r}-\mathbf{a}, \mathbf{a})}.
\end{aligned} \tag{3.62}$$

One also notes that these functions are related to each other according to the relation

$$f_{s-}(\mathbf{r}, \mathbf{a}) = f_{s+}(\mathbf{r} - \mathbf{a}, \mathbf{a}), \tag{3.63}$$

as can be expected. Finally, from the bulk limit of Eq. (3.61), one gets the relation between the solvent fugacity and the reservoir density in the form

$$\rho_{sb} = \Lambda_s \exp\left(E_s - Q^2 \left[v_0^b(0) - v_0^b(a) \right] \right). \tag{3.64}$$

After this lengthy gymnastics with the different terms in the model we are now ready to compute the variational equations. By taking the functional derivative of the grand potential (3.55) with respect to the trial potential $\phi_0(\mathbf{r})$, the equation of state determining the external potential follows as

$$\frac{k_B T}{e^2} \nabla_\mathbf{r} \varepsilon_0(\mathbf{r}) \nabla_\mathbf{r} \phi_0(\mathbf{r}) + \sum_i q_i \rho_i(\mathbf{r}) + Q \left[\rho_{s+}(\mathbf{r}) - \rho_{s-}(\mathbf{r}) \right] = -\sigma(\mathbf{r}). \tag{3.65}$$

Then, the functional derivative of the grand potential (3.55) with respect to the propagator $v_0(\mathbf{r}, \mathbf{r}')$ results in the following relation for the electrostatic kernel,

$$v_0^{-1}(\mathbf{r}, \mathbf{r}') = -\frac{k_B T}{e^2} \nabla_{\mathbf{r}} \varepsilon_0(\mathbf{r}) \nabla_{\mathbf{r}} \delta(\mathbf{r} - \mathbf{r}') + \sum_i q_i^2 \rho_i(\mathbf{r}) \delta(\mathbf{r} - \mathbf{r}')$$

$$+ Q^2 \int \frac{d\Omega}{4\pi} F[\mathbf{r}, \mathbf{r}', \mathbf{a}] \tag{3.66}$$

with

$$F[\mathbf{r}, \mathbf{r}', \mathbf{a}] \equiv f_{s+}(\mathbf{r}, \mathbf{a}) \left[\delta(\mathbf{r}' - \mathbf{r}) - \delta(\mathbf{r}' - \mathbf{r} - \mathbf{a}) \right]$$
$$+ f_{s-}(\mathbf{r}, \mathbf{a}) \left[\delta(\mathbf{r}' - \mathbf{r}) - \delta(\mathbf{r}' - \mathbf{r} + \mathbf{a}) \right]. \tag{3.67}$$

Using the definition of the Green's function formulated in Chap. 2, Eq. (2.117), we can invert the kernel (3.66) and finally obtain the variational equation for the electrostatic Green's function as

$$\delta(\mathbf{r} - \mathbf{r}') = \left[-\frac{k_B T}{e^2} \nabla_{\mathbf{r}} \varepsilon_0(\mathbf{r}) \nabla_{\mathbf{r}} + \sum_i q_i^2 \rho_i(\mathbf{r}) \right] v_0(\mathbf{r}, \mathbf{r}')$$

$$+ Q^2 \int \frac{d\Omega}{4\pi} \left\{ f_{s+}(\mathbf{r}, \mathbf{a}) \left[v_0(\mathbf{r}, \mathbf{r}') - v_0(\mathbf{r} + \mathbf{a}, \mathbf{r}') \right] \right.$$

$$\left. + f_{s-}(\mathbf{r}, \mathbf{a}) \left[v_0(\mathbf{r}, \mathbf{r}') - v_0(\mathbf{r} - \mathbf{a}, \mathbf{r}') \right] \right\}. \tag{3.68}$$

Equations (3.65) and (3.68) that incorporate the solvent charge structure generalize the implicit-solvent variational equations by Netz and Orland of Ref. [17] that we had discussed in the previous chapter. The first integro-differential Eq. (3.65) is a non-local Poisson-Boltzmann equation including charge fluctuation effects embodied in the ionic and dipolar self-energies [see Eqs. (3.59) and (3.61)–(3.62)]. In the absence of correlations, where these ionic and dipolar self-energies vanish, Eq. (3.65) reduces to the mean-field NLPB equation discussed in the previous chapter. The second Eq. (3.68) for the electrostatic Green's function is a *solvent-explicit Debye-Hückel equation* of non-local form. The non-local or integro-differential form of these equations is a consequence from the extended structure of the solvent molecules.

In the next step we will simplify the general Eqs. (3.65) and (3.68) for a planar geometry. The polar liquid and the solvated ions are confined to a slit nanopore with rigid interfaces located at $z = 0$ and $z = d$ (see Fig. 3.5). As we used to write in Chap. 2 for a slit geometry, the dielectric permittivity function can be written as

$$\varepsilon_0(\mathbf{r}) = \varepsilon_0(z) = \varepsilon_0 \theta(z) \theta(d - z) + \varepsilon_m \theta(-z) \theta(z - d) \tag{3.69}$$

with the vacuum permittivity $\varepsilon_0 = 1$ and the membrane permittivity ε_m. The confinement is imposed by the wall potential $W_i(\mathbf{r}) = W_i(z) = 0$ if $0 \leq z \leq d$ and $W_i(z) = \infty$ otherwise. Moreover, in terms of the projection of the dipolar alignment on the z axis one has

$$a_z = a \cos\theta, \tag{3.70}$$

where θ is the angle between the dipole and the z axis (see Fig. 3.3). The dipolar wall potential imposing the solvent confinement is given by $W_s(\mathbf{r}, \mathbf{a}) = W_s(z, a_z) = 0$ if $0 \leq z \leq d$ and $0 \leq z + a_z \leq d$, and $W_s(z, a_z) = \infty$ otherwise.

Exploiting the translational symmetry in the (x, y)-plane within the planar geometry, one can expand the Green's function in Fourier space as we did already in Chap. 2 in Eq. (2.123). To simplify the notation from now on we will omit the k-dependence of the Fourier-transformed Green's function. Carrying out in Eq. (2.123) the integral over the angle $\theta_{\mathbf{k}}$ in the reciprocal plane, one obtains

$$v_0(\mathbf{r}, \mathbf{r}') = \int_0^\Lambda \frac{dk\,k}{2\pi} J_0\left[k|\mathbf{r}_\parallel - \mathbf{r}'_\parallel|\right] v_0(z, z'), \tag{3.71}$$

with the ultraviolet (UV) cut-off Λ and the *Bessel function of the first kind* $J_0(x)$. Inserting the expression of Eq. (3.71) into Eq. (3.57), the dipolar self-energy follows in the form

$$v_d(z, a_z) = \int_0^\Lambda \frac{dk\,k}{2\pi} \left[v_0(z, z) + v_0(z + a_z, z + a_z) - 2v_0(z, z + a_z) J_0\left[k|a_\parallel|\right]\right], \tag{3.72}$$

where we defined the projection of the dipolar vector $\mathbf{a}$ onto the (x, y)-plane as $a_\parallel = a \sin \theta$. The amplitude of the parallel component $a_\parallel$ in Eq. (3.72) is related to the perpendicular component a_z through the relation $|a_\parallel| = \sqrt{a^2 - a_z^2}$.

Using in Eqs. (3.59) and (3.61)–(3.62) the bulk relations (3.60) and (3.64) between the particle densities and fugacities, and defining the renormalized ionic and dipolar self-energies

$$\delta v_i(z) = v_0(z, z) - v_0^b(0) \tag{3.73}$$

$$\delta v_d(z, a_z) = v_d(z, a_z) - 2v_0^b(0) + 2v_0^b(a), \tag{3.74}$$

the ion and solvent number densities follow in the form

$$\rho_i(z) = \rho_{ib} e^{-q_i \phi_0(z)} e^{-\frac{q_i^2}{2} \delta v_i(z) - W_i(z)} \tag{3.75}$$

$$\rho_{s\pm}(z) = \int_{a_1(z)}^{a_2(z)} \frac{da_z}{2a} f_{s\pm}(z, a_z), \tag{3.76}$$

with the solvent molecular charge densities at fixed orientation

$$f_{s\pm}(z, a_z) = \rho_{sb} e^{-\frac{Q^2}{2} \delta v_d(z, a_z)} e^{\pm Q[\phi_0(z + a_z) - \phi_0(z)]}. \tag{3.77}$$

In Eq. (3.76), we introduced the integral boundaries taking into account the impenetrability of the interfaces,

$$a_1(z) = -\min(a, z) \tag{3.78}$$

$$a_2(z) = \min(a, d - z). \tag{3.79}$$

Furthermore, in passing from Eqs. (3.61)–(3.76), we performed the change of variable $\theta \rightarrow a_z$ in the integral over the dipole rotations. Substituting now the solvent density (3.76) into the variational NLPB equation, Eq. (3.65), and inserting the Fourier expansion of the Green's function (2.123) and the solvent density function (3.77) into the second variational Eq. (3.68), the electrostatic self-consistent equations take the simpler form for $0 \leq z \leq d$

$$\frac{k_B T}{e^2} \partial_z \varepsilon_0(z) \partial_z \phi_0(z) + \sum_i q_i \rho_i(z)$$

$$+ 2Q \rho_{sb} \int\limits_{a_1(z)}^{a_2(z)} \frac{da_z}{2a} \sinh\left[Q\phi_0(z + a_z) - Q\phi_0(z)\right] e^{-\frac{Q^2}{2}\delta v_d(z, a_z)} = -\sigma(z) \tag{3.80}$$

and

$$-\frac{k_B T}{e^2} \left[\partial_z \varepsilon_0(z) \partial_z - \varepsilon_0(z) p^2(z)\right] v_0(z, z')$$

$$+ 2Q^2 \rho_{sb} \int\limits_{a_1(z)}^{a_2(z)} \frac{da_z}{2a} \cosh\left[Q\phi_0(z + a_z) - Q\phi_0(z)\right] e^{-\frac{Q^2}{2}\delta v_d(z, a_z)}$$

$$\times \left[v_0(z, z') - v_0(z + a_z, z') J_0(k|a_\parallel|)\right] = \delta(z - z'). \tag{3.81}$$

In Eq. (3.81), we have introduced the auxiliary function

$$p(z) = \sqrt{k^2 + \kappa_i^2(z)} \tag{3.82}$$

with the *ion screening function*

$$\kappa_i^2(z) = \frac{e^2}{\varepsilon_0(z) k_B T} \sum_i q_i^2 \rho_i(z). \tag{3.83}$$

3.5   The Dielectric Permittivity $\varepsilon(k)$: Mean-Field Versus Variational Models

In a first step let us establish the link between the variational model of this section and the mean-field case studied in the previous section. For this we let the distance d between the two plates tend to infinity so that we are left with a single plate at $z = 0$. We then solve the variational equation for the electrostatic potential, Eq. (3.80), approximating the ionic and dipolar self energies by the local Debye-Hückel potential; details can be found in Ref. [14].

Figure 3.6a shows the result for the dielectric permittivity as computed from the electrostatic potential profiles $E(z) = \phi'_0(z)$ obtained from the solution of Eq. (3.80) [18] via

$$E(z) = \frac{e^2 \sigma_s}{k_B T \varepsilon_{eff}(z)}. \tag{3.84}$$

In order to extend this relation beyond the mean-field case, we note that by integrating Eq. (3.80) from the interface to a given point z in the liquid and using relation (3.84), the inverse effective permittivity can be expressed in terms of the cumulative polarization charge between the surface and the point z,

$$\frac{1}{\varepsilon_{eff}(z)} = 1 - \frac{1}{\sigma_s} \int\limits_0^z dz' \rho_{sc}(z'), \tag{3.85}$$

with the solvent charge density given by

$$\rho_{sc}(z) = Q \left[\rho_{s+}(z) - \rho_{s-}(z) \right]. \tag{3.86}$$

Equation (3.85) was first derived in Ref. [18] for the mean-field NLPB model. The relation states that the trend of the effective permittivity is reversed at the points where the polarization charge density changes its sign, see Fig. 3.6b. The oscillations of the permittivity result from the presence of successive hydration layers with alternating net charge at the interface. For all membrane permittivities, the extrema of the effective permittivity correspond to the boundary between two neighboring hydration shells of opposite charge.

We have restricted ourselves to the linear response regime and eliminated ionic screening effects, considering a very weak surface charge density $\sigma_s = 10^{-6}$ e nm^{-2} and a dilute salt with bulk concentration $\rho_{ib} = 10^{-5}$ M. The numerical result for the dielectric permittivity profile is shown in Fig. 3.6a for membrane permittivities of $\varepsilon_m = 1$ and $\varepsilon_m = 50$. At the mean-field level, or equivalently for the dielectrically homogeneous system with $\varepsilon_m = \varepsilon_w$ where one is left exclusively with non-local dielectric response effects, it is seen that the interface is characterized by a dielectric void followed by pronounced oscillations of the effective permittivity function around the bulk permittivity; this is the black dashed curve for the mean-field NLPB-model.

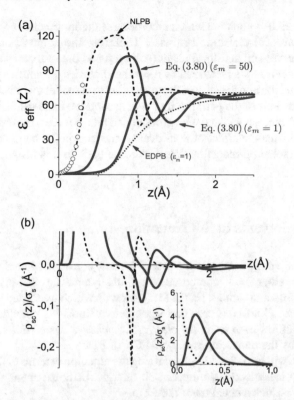

Fig. 3.6 a The effective dielectric permittivity profile for a solvent with bulk density $\rho_{sb} = 50.8$ M and permittivity $\varepsilon_w = 71$, in contact with a planar interface with surface charge $\sigma_s = 10^{-6}$ e nm^{-2}, z is measured in Å. The salt concentration is $\rho_{ib} = 10^{-5}$ M. The graph shows the nonlocal mean-field result (NLPB, dashed black curve), reached in the limit $\varepsilon_m = \varepsilon_w$. The effect of correlations associated with the dielectric inhomogeneity between the solvent and the membrane are shown by the solid curves. The blue and red curves are obtained from Eq. (3.80), for two different values of the dielectric constant of the pore walls. The blue dashed curve shows the permittivity profile of the point-dipole variational model (EDPB), hence a model without the nonlocal effect. Also shown is the constant Poisson-Boltzmann permittivity as the reference line at $\varepsilon_w = 77$. The open circles in **a** correspond to the asymptotic behaviour $\varepsilon(z) = e^{\kappa_s z/\sqrt{2}}$. **b** Corresponding rescaled polarization charge density. For the discussion, see text. Reproduced from [14], with the permission of AIP Publishing

Furthermore, in Fig. 3.6 one observes that for membranes with low permittivity $\varepsilon_m < \varepsilon_w$, electrostatic correlations increase the size of the *dielectric void* close to the charged surface, shift the permittivity curves towards larger distances from the interface, and reduce the overall amplitude of the dielectric oscillations; for e.g. for a value of $\varepsilon_m = 50$ the resulting curve for $\varepsilon_{eff}(z)$ interpolates between the mean-field and variational result for small ε_m.

Figure 3.6a also shows the dielectric permittivity profile of the variational point-dipole model developed in Ref. [16] as a blue dashed curve. It is obtained from the solution of the EDPB equation which incorporates the interfacial solvent exclusion but not the non-local dielectric response. Therefore the comparison of the local (dashed blue curve) and non-local result (solid red and blue curves) shows that both variational models are characterized by a similar dielectric permittivity reduction at the interface, and the non-locality of the present solvent model manifests itself by the oscillatory behaviour of the permittivity curve around the local result. Thus, for low dielectric membranes associated with pronounced image forces, the resulting solvent depletion dominates the effect of non-locality and brings the main contribution to the interfacial dielectric reduction. We will come back to this insight at the end of the chapter.

3.6 Dilute Solvents in Slit Nanopores

As the next detailed application of the theory we now consider a polar liquid confined within a neutral slit, $\sigma_s = 0$, composed of a dilute symmetric electrolyte containing monovalent anions and cations, i.e. the (1:1) salt with which we started into this book. According to Eq. (3.80), this case corresponds to a vanishing potential, $\phi_0(z) = 0$. This example allows us to study the *electrostatic charge correlations* in this 'nano-slit' governed by the remaining variational Eq. (3.81).

The solution will be obtained by linearizing the equation in terms of the propagator or, equivalently, in neglecting the dipolar self-energies in the exponentials. This yields a dipolar Debye-Hückel equation of the form

$$
-\frac{k_B T}{e^2} \left[\partial_z \varepsilon_0(z) \partial_z - \varepsilon_0(z) p_b^2 \right] v_0(z, z')
$$

$$
+2 Q^2 \rho_{sb} \int\limits_{a_1(z)}^{a_2(z)} \frac{\mathrm{d}a_z}{2a} \left\{ v_0(z, z') - v_0(z + a_z, z') J_0(k|a_\parallel|) \right\} = \delta(z - z'), \qquad (3.87)
$$

with $p_b = \sqrt{k^2 + \kappa_{ib}^2}$ and the screening parameter

$$
\kappa_{ib}^2 = \frac{e^2}{\varepsilon_0 k_B T} \sum_i q_i^2 \rho_{ib} \theta(z) \theta(d - z). \qquad (3.88)
$$

The usual Debye-Hückel equation of the dielectric continuum description follows from the point-dipole limit of Eq. (3.87). By expanding Eq. (3.87) up to quadratic order in the solvent molecular size a, relaxing the rotational penalty for dipoles by setting $a_1(z) = -a$ and $a_2(z) = a$, and carrying out the integral over the dipole rotations, one ends up with the usual Debye-Hückel equation in a planar geometry, which reads as

$$\left[\partial_z \varepsilon(z)\partial_z - \varepsilon(z)\left(k^2 + \kappa_{DH}^2\right)\right]v_{DH}(z,z') = -\frac{e^2}{k_B T}\delta(z-z'), \quad (3.89)$$

with the dielectric permittivity function given by the expression

$$\varepsilon(z) = \varepsilon_0\left[\theta(-z) + \theta(z-d)\right] + \varepsilon_w\theta(z)\theta(d-z) \quad (3.90)$$

and the Debye-Hückel screening parameter defined as

$$\kappa_{DH}^2 \equiv \frac{e^2}{\varepsilon_w k_B T}\sum_i q_i^2 \rho_{ib}. \quad (3.91)$$

We now explain the procedure to rephrase Eq. (3.87) for dilute solvents such that it becomes amenable to numerical treatment. The latter we do not present is this text; it can be found detailed in the appendix of the original Ref. [14]. We first define the second term on the left-hand side of the equation as the auxiliary function

$$F(z,z') \equiv 2Q^2 \rho_{sb} \int_{a_1(z)}^{a_2(z)} \frac{da_z}{2a}\left\{v_0(z,z') - v_0(z+a_z,z')J_0(k|a_\parallel|)\right\}, \quad (3.92)$$

and the reference kernel corresponding to the Debye-Hückel kernel in vacuum as

$$v_{ref}^{-1}(z,z') \equiv -\frac{k_B T}{e^2}\left[\partial_z \varepsilon_0(z)\partial_z - \varepsilon_0(z)p_b^2\right]\delta(z-z'). \quad (3.93)$$

Making use of Eq. (2.117), the relation (3.87) can be formally inverted as

$$v_0(z,z') = v_{ref}(z,z') - \int_0^d dz_1 v_{ref}(z,z_1)F(z_1,z'). \quad (3.94)$$

The derivation of the reference Debye-Hückel potential in vacuum is similar to the computation of the same potential in the dielectric continuum solvent, solution of Eq. (3.89) (see e.g. Ref. [19] for details). It is given by the sum of a bulk and an interfacial part,

$$v_{ref}(z,z') = v_{ref,b}(z,z') + \delta v_{ref}(z,z'), \quad (3.95)$$

where the homogeneous part is given by

$$v_{ref,b}(z,z') = \frac{2\pi\ell_B}{p_b}e^{-p_b|z-z'|}, \quad (3.96)$$

and the contribution from the solvent confinement in the slit is given by

$$\delta v_{ref}(z, z') = \frac{2\pi \ell_B}{p_b} \frac{\Delta}{1 - \Delta^2 e^{-2p_b d}}$$
$$\times \left\{ e^{-p_b(z+z')} + e^{-p_b(2d-z-z')} + 2\Delta e^{-2p_b d} \cosh\left(p|z - z'|\right) \right\}, \quad (3.97)$$

with the auxiliary function

$$\Delta = \frac{p_b - k}{p_b + k}. \quad (3.98)$$

Equation (3.94) allows to compute the non-local potential $v_0(z, z')$ by numerical iteration around the reference potential (3.95). It is useful to note that the Debye-Hückel potential $v_{DH}(z, z')$ in the dielectric continuum solvent can be recovered from Eq. (3.95) if one replaces the coefficients ℓ_B, p_b, and Δ in Eqs. (3.96)–(3.97) respectively by the Bjerrum length in the solvent

$$\ell_w = \frac{\ell_B}{\varepsilon_w}, \quad (3.99)$$

the screening function

$$\bar{p} = \sqrt{k^2 + \kappa_{DH}^2}, \quad (3.100)$$

and the dielectric jump function

$$\bar{\Delta} = \frac{\varepsilon_w \bar{p} - \varepsilon_m k}{\varepsilon_w \bar{p} + \varepsilon_m k}. \quad (3.101)$$

For the numerical solution of Eq. (3.87) the bulk salt density is set to $\rho_{ib} = 10^{-6}$ M. Since our solvent model is made up of finite-size dipoles, Eqs. (3.80)–(3.81) and Eq. (3.87) do not show any UV-divergences. However, in order to simplify the numerical task, the integrals in Fourier space were computed with a finite ultraviolet (UV) cut-off $\Lambda = 1000/\ell_B$. The self-energies of the ions, as defined in Eq. (3.73), are shown in Fig. 3.7 for dilute solvents with a density of 0.1 M. We also display as dashed curves the local image charge potentials corresponding to the solution of Eq. (3.89) with the finite cut-off $\Lambda = 1000/\ell_B$.

One sees that for the given solvent concentration, the self-energies of the ions in the dipolar and dielectric continuum liquids are fairly close to each other, the non-local potential being only slightly higher than the Debye-Hückel potential close to the slit walls. This result shows that the explicit consideration of solvent interactions naturally results in the *image-charge forces* usually obtained by imposing the dielectric jump between the membrane medium and the solvent.

Fig. 3.7 a Ionic
self-energies for different
pore sizes at the solvent
concentration $\rho_{sb} = 0.1$ M;
b Ion densities for pores of
size $d = 20$ Å (main plot)
and $d = 5$ Å (inset). Solid
and dashed curves
correspond, respectively, to
the solutions obtained from
the non-local Eq. (3.87) and
the dielectric continuum
Eq. (3.89). The bulk ion
concentration is
$\rho_{ib} = 10^{-6}$ M, and the slit is
neutral ($\sigma_s = 0.0$ e nm^{-2}).
For the discussion, see text.
Reproduced from [14], with
the permission of AIP
publishing

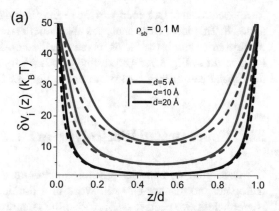

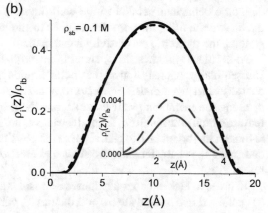

We can also consider the effect of the confinement on the ionic self-energies
and densities. The reduction of the pore size in Fig. 3.7a and the corresponding
increase of the ionic self-energies goes qualitatively hand-in-hand with an increase
of the bulk solvent concentration, that is, it amplifies the energetic barrier for charge
penetration. One notes that the difference between the local and the non-local self-
energy becomes relevant for pores of sub-nanometer size, i.e. if the confinement
scale becomes comparable to the size of solvent molecules, see Fig. 3.7b.

Within the present explicit solvent theory, the energetic barrier for ionic penetration into the pore is composed of two contributions. First, since the pore dielectric
permittivity is larger than the membrane permittivity, mobile charges experience a
stronger dielectric screening in the mid-pore region than close to the membrane.
Hence, ions feel a repulsive force excluding them from the pore wall, which is the
well-known *image-charge effect* already present in the dielectric continuum electrostatics. The second contribution is the *Born energy difference* between the pore
and the bulk reservoir. Due to the confinement in the pore, the solvent density and
dielectric permittivity are lower than in the bulk reservoir. As a result, ions possess a

lower electrostatic free energy in the reservoir and this favors their rejection from the nanoslit. This ionic Born energy is absent in the dielectric continuum formulation of electrostatics and is the factor increasing the non-local ionic self energies above the local result in Fig. 3.7a. The results discussed in this part still neglect the effect of the dipolar self-energy $\delta v_d(z)$ present in Eq. (3.81).

3.7   Concentrated Electrolytes in Slit Nanopores

In this section we consider physiological concentrations of electrolytes inside slit nanopores, so that we can still work in a dominantly one-dimensional geometry. Nevertheless, we need to characterize the anisotropy of this situation, in which the dielectric behaviour parallel to the wall deviates from the behaviour transverse to the electrolyte filing the pore. Remember from the beginning of this chapter that, in general, the dielectric permittivity is a tensor.

Molecular Dynamics (MD) simulations with explicit solvent by Balleneger and Hansen had previously found that polar liquids confined in nanopores are indeed characterized by a *dielectric anisotropy* associated with a transverse permittivity $\varepsilon_\parallel$ along the membrane wall that exceeds the perpendicular component $\varepsilon_\perp$ [20], a feature which is absent from the dielectric continuum formulation of electrostatics. It is thus an important task to characterize the physics behind these effects, and, e.g., to determine the characteristic pore sizes and electrolyte densities where they become relevant.

In order to tackle these questions we consider again a salt free solvent ($\rho_{ib} = 0.0$ M) confined in a neutral slit $\sigma_s = 0.0$ e nm^{-2}, which results in a vanishing external potential $\phi_0(\mathbf{r}) = 0$. The iterative numerical method employed in the previous chapter fails for such high solvent density. Instead we will introduce a restricted local *self-consistent approach* which is general enough to capture these effects.

This restricted self-consistent approach consists in computing the variational grand potential (3.55) with a *trial ansatz* that solves a *dielectrically anisotropic Laplace equation* [14]

$$\left[\nabla_{\mathbf{r}_\parallel} \varepsilon_\parallel(z) \nabla_{\mathbf{r}_\parallel} + \partial_z \varepsilon_\parallel(z) \partial_z\right] v_0(\mathbf{r}, \mathbf{r}') = -\frac{e^2}{k_B T} \delta(\mathbf{r} - \mathbf{r}'), \qquad (3.102)$$

with the dielectric permittivity functions parallel and perpendicular to the pore walls introduced by setting

$$\varepsilon_{\parallel,\perp}(z) = \varepsilon_{\parallel,\perp} \theta(z)\theta(d - z) + \varepsilon_m \left[\theta(-z) + \theta(z - d)\right]. \qquad (3.103)$$

We need to solve Eq. (3.102) in order to compute the grand potential as a function of the permittivity components $\varepsilon_\parallel$ and $\varepsilon_\perp$—these will be considered as variational parameters whose numerical values will be obtained from the minimization of the variational grand potential. We will see that the grand potential can be written as consisting of three contributions

$$\Omega_v = \Omega_0 + \Omega_c + \Omega_d, \tag{3.104}$$

which correspond to a *van der Waals part* Ω_0, a *correction term* Ω_c and the contribution from the *dipolar density* Ω_d, whose computation we will describe in detail.

The first task, however, is to solve Eq. (3.102).

3.8 The Green Function of the Dielectrically Anisotropic Laplace Equation

We start by using the Fourier expansion (2.123) from the previous chapter, and insert it into Eq. (3.102).The equation then takes the one-dimensional form

$$\left[-\partial_z \varepsilon_\perp(z)\partial_z + \varepsilon_\parallel(z)k^2\right] v_0(z, z') = \frac{e^2}{k_B T}\delta(z - z'), \tag{3.105}$$

with the dielectric permittivity components $\varepsilon_{\parallel,\perp}(z)$ from Eq. (3.103).

We will need the solution of Eq. (3.105) exclusively for the charges located in the slit, i.e. for $0 \leq z' \leq d$. The general solution of Eq. (3.105) is then given by the superposition of exponentials, multiplied with the Heaviside functions that take care of the confinement conditions, in a similar fashion as we built the solution for the polymer adsorption problem in Chap. 2. We have

$$\begin{aligned}
v_0(z, z') &= C_1 e^{kz}\theta(-z) + C_2 e^{-kz}\theta(z - d) \\
&\quad + \left[C_3 e^{\Gamma kz} + C_4 e^{-\Gamma kz}\right]\theta(z' - z)\theta(z)\theta(d - z) \\
&\quad + \left[C_5 e^{\Gamma kz} + C_6 e^{-\Gamma kz}\right]\theta(z - z')\theta(z)\theta(d - z), \tag{3.106}
\end{aligned}$$

where we introduced the *anisotropy coefficient*

$$\Gamma = \sqrt{\frac{\varepsilon_\parallel}{\varepsilon_\perp}}. \tag{3.107}$$

The constants C_i with $i = 1, \ldots, 6$ in Eq. (3.106) are, in the same way as in the polymer adsorption problem of Chap. 2, obtained from the continuity of the electrostatic potential $v_0(z, z')$ and the displacement field $\varepsilon_\perp(z)\partial_z v_0(z, z')$ at the slit walls and at the location of the charge source $z = z'$. One obtains the Fourier-transformed propagator in the form

$$v_0(z, z') = v_{0b}(z, z') + \delta v_0(z, z'), \tag{3.108}$$

with the bulk part

$$v_{0b}(z, z') = \frac{2\pi \ell_B}{k\sqrt{\varepsilon_\perp \varepsilon_\parallel}} e^{-\Gamma k |z-z'|} \tag{3.109}$$

and the dielectric part

$$\delta v_0(z, z') = \frac{2\pi \ell_B}{k\sqrt{\varepsilon_\perp \varepsilon_\parallel}} \frac{\Delta_\Gamma}{1 - \Delta_\Gamma^2 e^{-2\Gamma kd}}$$
$$\times \left\{ e^{-\Gamma k(z+z')} + e^{-\Gamma k(2d-z-z')} + 2\Delta_\Gamma e^{-2\Gamma kd} \cosh\left(\Gamma k |z - z'|\right) \right\}, \tag{3.110}$$

where we defined the dielectric discontinuity function

$$\Delta_\Gamma = \frac{\sqrt{\varepsilon_\perp \varepsilon_\parallel} - \varepsilon_m}{\sqrt{\varepsilon_\perp \varepsilon_\parallel} + \varepsilon_m}. \tag{3.111}$$

3.9 The Three Contributions to the Variational Grand Potential Ω_v

We now compute the three contributions to the grand potential, Eq. (3.104).

3.9.1 The Anisotropic van der Waals Free Energy

The *van der Waals part* Ω_0 of the variational grand potential is given by the expression

$$\Omega_0 = -\ln \int \mathcal{D}\phi \; e^{-\frac{k_B T}{2e^2} \int d\mathbf{r} \left\{ \varepsilon_\parallel(z) \left[(\partial_x \phi)^2 + (\partial_y \phi)^2 \right] + \varepsilon_\perp(z)(\partial_z \phi)^2 \right\}}. \tag{3.112}$$

In order to evaluate this functional integral, the charging procedure introduced in Ref. [19] for the computation of the isotropic van der Waals-energy will be used. This procedure allows us to interpolate between different expressions of the dielectric functions. Here it consists in the introduction of two auxiliary permittivity functions with the *charging parameters* ξ and η,

$$\varepsilon_\xi(z) = \varepsilon_\perp(z) + \xi \left[\varepsilon_\parallel(z) - \varepsilon_\perp(z) \right] \tag{3.113}$$

$$\varepsilon_\eta(z) = \varepsilon_m + \eta \left[\varepsilon_\perp(z) - \varepsilon_m \right]. \tag{3.114}$$

Expressing Eq. (3.112) in terms of two auxiliary integrals over ξ and η, one finds

$$\Omega_0 = -\int_0^1 d\xi \frac{d}{d\xi} \ln \int \mathcal{D}\phi \, e^{-\frac{k_BT}{2e^2} \int d\mathbf{r} \left\{ \varepsilon_\xi(z) \left[(\partial_x\phi)^2 + (\partial_y\phi)^2 \right] + \varepsilon_\perp(z)(\partial_z\phi)^2 \right\}}$$

$$-\int_0^1 d\eta \frac{d}{d\eta} \ln \int \mathcal{D}\phi \, e^{-\frac{k_BT}{2e^2} \int d\mathbf{r}\, \varepsilon_\eta(z)(\nabla_\mathbf{r}\phi)^2} - \ln \int \mathcal{D}\phi \, e^{-\frac{k_BT}{2e^2} \int d\mathbf{r}\, \varepsilon_m (\nabla\phi)^2}. \quad (3.115)$$

The third integral on the right-hand side of Eq. (3.115) is the free energy of a bulk medium with dielectric permittivity ε_m. Since this contribution is independent of the variational parameters $\varepsilon_\perp$ and $\varepsilon_\parallel$ and the slit size d, it can be dropped in the following derivation. Evaluating the derivatives acting on the functional integral in Eq. (3.115), the free energy takes the form

$$\Omega_0 = \frac{k_BT}{2e^2} \int d\mathbf{r} \left\{ \int_0^1 d\xi \left[\varepsilon_\parallel(z) - \varepsilon_\perp(z) \right] \langle (\nabla_{\mathbf{r}_\parallel}\phi)^2 \rangle_{\varepsilon_\parallel(z) \to \varepsilon_\xi(z)} \right.$$

$$\left. + \int_0^1 d\eta \left[\varepsilon_\perp(z) - \varepsilon_m \right] \langle (\nabla_\mathbf{r}\phi)^2 \rangle_{\varepsilon_{\parallel,\perp}(z) \to \varepsilon_{\eta(z)}} \right\}. \quad (3.116)$$

In Eq. (3.116), the subscripts of the brackets mean that the averages should be evaluated with the electrostatic Green's function (3.108) by replacing the dielectric permittivity profiles of the latter with the auxiliary permittivity functions (3.113) and (3.114),

$$v_0^\xi(\mathbf{r}, \mathbf{r}') = v_0 \left[\mathbf{r}, \mathbf{r}'; \varepsilon_\parallel(z) \to \varepsilon_\xi(z) \right] \quad (3.117)$$

$$v_0^\eta(\mathbf{r}, \mathbf{r}') = v_0 \left[\mathbf{r}, \mathbf{r}'; \varepsilon_{\parallel,\perp}(z) \to \varepsilon_\eta(z) \right]. \quad (3.118)$$

Evaluating the field-theoretic averages, the free energy (3.116) can be expressed in terms of the propagators (3.117) and (3.118) as

$$\Omega_0 = \frac{k_BT}{2e^2} \int d\mathbf{r} \left\{ \int_0^1 d\xi \left[\varepsilon_\parallel(z) - \varepsilon_\perp(z) \right] \nabla_{\mathbf{r}_\parallel} \cdot \nabla_{\mathbf{r}'_\parallel} v_0^\xi(\mathbf{r}, \mathbf{r}') \right.$$

$$\left. + \int_0^1 d\eta \left[\varepsilon_\perp(z) - \varepsilon_m \right] \nabla_\mathbf{r} \cdot \nabla_{\mathbf{r}'} v_0^\eta(\mathbf{r}, \mathbf{r}') \right\} \Bigg|_{\mathbf{r}' \to \mathbf{r}} \quad (3.119)$$

hence

$$
\Omega_0 = S \frac{k_B T}{4\pi e^2} \int\limits_0^d dz \int\limits_0^\Lambda dk \left\{ (\varepsilon_\| - \varepsilon_\perp) \int\limits_0^1 d\xi \, k^2 v_0^\xi(z, z') \right.
$$

$$
\left. + (\varepsilon_\perp - \varepsilon_m) \int\limits_0^1 d\eta \left[k^2 + \partial_z \partial_{z'} \right] v_0^\eta(z, z') \right\} \Bigg|_{z' \to z} \tag{3.120}
$$

where S stands for the lateral surface of the membrane. Equation (3.120) follows from Eq. (3.119) after substituting the Fourier expansion of the electrostatic propagator Eq. (2.123). In order to compute the integrals, we need to make use of Eq. (3.108). The two derivatives acting on the bulk part of the Green's function Eq. (3.109) in the last term of Eq. (3.120) yield a delta function evaluated at zero which is finite since the UV modes are regularized with a cut-off.

Carrying out the integrals over the pore size and the auxiliary parameters in Eq. (3.120) one obtains the anisotropic van der Waals part of the grand potential:

$$
\Omega_0 = \frac{Sd\Lambda^3}{12\pi} \frac{\sqrt{\varepsilon_\|} - \sqrt{\varepsilon_\perp}}{\sqrt{\varepsilon_\perp}} + \frac{Sd\Lambda^3}{16\pi} \ln \frac{\varepsilon_\perp}{\varepsilon_m}
$$

$$
+ \frac{S\Lambda^2}{8\pi} \ln \frac{(\sqrt{\varepsilon_\| \varepsilon_\perp} + \varepsilon_m)^2}{4\varepsilon_m \sqrt{\varepsilon_\| \varepsilon_\perp}} + \frac{S}{4\pi} \int\limits_0^\Lambda dk\, k \ln \left(1 - \Delta_\Gamma^2 e^{-2\Gamma kd} \right), \tag{3.121}
$$

where S is the lateral surface of the membrane with dielectric anisotropy and discontinuity functions Γ and Δ_Γ given as before. The first two terms on the right-hand side of Eq. (3.121) correspond to the electrostatic energy of a bulk medium of volume $V_p = Sd$. The third term is the surface tension of two decoupled interfaces separating the pore walls and the solvent media with permittivities ε_m and $\sqrt{\varepsilon_\| \varepsilon_\perp}$, respectively. Finally, the fourth term is the interaction energy of these interfaces located at $z = 0$ and $z = d$.

3.9.2   The Correction Term Ω_c

The correction term Ω_c in Eq. (3.104) is given by the expression

$$
\Omega_c = S \frac{k_B T}{2e^2} \int d\mathbf{r} \left\{ \left[\varepsilon_0(z) - \varepsilon_\|(z) \right] \nabla_{\mathbf{r}_\|} \cdot \nabla_{\mathbf{r}'_\|} \right.
$$

$$
\left. + \left[\varepsilon_0(z) - \varepsilon_\perp(z) \right] \partial_z \partial_{z'} \right\} v_0(\mathbf{r}, \mathbf{r}') \Big|_{\mathbf{r}' \to \mathbf{r}}. \tag{3.122}
$$

Inserting the anisotropic propagator of Eq. (3.108) into this equation, taking into account the equality $\varepsilon_m = \varepsilon_0$, and carrying out the spatial integrals over the slit width, one is led to

$$\Omega_c = \frac{Sd\Lambda^3}{48\pi}\left(2\frac{\varepsilon_0 - \varepsilon_\parallel}{\sqrt{\varepsilon_\perp \varepsilon_\parallel}} + \Gamma\frac{\varepsilon_0 - \varepsilon_\perp}{\sqrt{\varepsilon_\perp \varepsilon_\parallel}}\right) + \frac{S\Lambda^2}{16\pi}\frac{\Delta_\Gamma}{\varepsilon_\parallel}\left[\varepsilon_0 - \varepsilon_\parallel + \Gamma^2(\varepsilon_0 - \varepsilon_\perp)\right]$$

$$+\frac{S\Delta_\Gamma}{8\pi\varepsilon_\parallel}\int_0^\Lambda \frac{dk k e^{-2\Gamma k d}}{1 - \Delta_\Gamma^2 e^{-2\Gamma k d}}$$

$$\times\left\{(\varepsilon_0 - \varepsilon_\parallel)(\Delta_\Gamma^2 + 2\Delta_\Gamma\Gamma k d - 1) + (\varepsilon_0 - \varepsilon_\perp)(\Delta_\Gamma^2 - 2\Delta_\Gamma\Gamma k d - 1)\right\}.$$

(3.123)

3.9.3 The Dipolar Contribution to the Grand Potential

The computation of the contribution from the dipole density to the grand potential (3.104) is more involved due to the finite length of the dipole molecule confined to the slit pore. After evaluating the field-theoretic average in Eq. (3.55), one finds for the contribution from the solvent molecules the expression

$$\Omega_d = -S\Lambda_s \int_0^d dz \int_{a_1(z)}^{a_2(z)} \frac{da_z}{2a} e^{E_s - \frac{\varrho^2}{2}v_d(a,a_z)}.$$

(3.124)

As before for the van der Waals term, the *dipolar self-energy* in Eq. (3.124) is composed of a bulk and a surface contribution, $v_d(z, a_z) = v_{db}(a_z) + v_{ds}(z, a_z)$, where the bulk part corresponding to the *Born energy* of solvent molecules is

$$v_{db}(a_z) = \frac{2\ell_B \varepsilon_0}{\sqrt{\varepsilon_\parallel \varepsilon_\perp}}\int_0^\Lambda dk\left[1 - e^{-\Gamma k |a_z|}J_0(ka_\parallel)\right],$$

(3.125)

and the inhomogeneous part that accounts for the confinement reads as

$$v_{ds}(z, a_z) = \int_0^\Lambda \frac{dk k}{2\pi}\left\{\delta v_0(z, z) + \delta v_0(z + a_z, z + a_z)\right.$$

$$\left.-2\delta v_0(z, z + a_z)J_0(ka_\parallel)\right\},$$

(3.126)

with the potential $\delta v_0(z, z')$ given by Eq. (3.110).

In order to obtain the relation between the solvent fugacity and the reservoir density, we have to compute the grand potential (3.104) for a bulk solvent. In the bulk limit $d \to \infty$ and $\Delta_\Gamma = 0$, the surface contribution (3.126) naturally vanishes.

Then, since we chose a local form (3.102) for the reference electrostatic kernel, the bulk self-energy (3.125) has to be expanded in the point-dipole limit in order to recover the cut-off free bulk solution for the effective permittivities. Taking the point-dipole limit of Eq. (3.125) that consists of its Taylor expansion up to the order $O(a^2)$, one gets the result

$$v_{db}(a_z) = \frac{\ell_B \Lambda^3 \varepsilon_0}{6\sqrt{\varepsilon_\parallel \varepsilon_\perp}} \left(a_\parallel^2 + \Gamma a_z^2 \right). \tag{3.127}$$

In the same bulk limit $V = Sd \to \infty$, the grand potential per volume $f_b = \Omega_v / V$ reads

$$f_b = \frac{\Lambda^3}{12\pi}\Gamma + \frac{\Lambda^3}{16\pi} \ln \frac{\varepsilon_\perp}{\varepsilon_m} + \frac{\Lambda^3}{48\pi} \left[\frac{2(\varepsilon_0 - \varepsilon_\parallel)}{\sqrt{\varepsilon_\parallel \varepsilon_\perp}} + \frac{\varepsilon_0}{\varepsilon_\perp} \right] - \Lambda_s \frac{\sqrt{\pi}}{2} \frac{\mathrm{erf}(u)}{u} e^{-\alpha} \tag{3.128}$$

with the coefficients

$$\alpha = \frac{Q^2 \Lambda^3 \ell_B a^2 \varepsilon_0}{12\sqrt{\varepsilon_\parallel \varepsilon_\perp}}, \quad u = \sqrt{\alpha(\Gamma - 1)}. \tag{3.129}$$

From the numerical minimization of the bulk grand potential (3.128) with respect to $\varepsilon_\parallel$ and $\varepsilon_\perp$, we find that the permittivity components are given by $\varepsilon_\parallel = \varepsilon_\perp = \varepsilon_w$ in agreement with the result of Ref. [18]. Substituting this solution into Eq. (3.128), and evaluating the solvent density with the thermodynamic relation

$$\rho_{sb} = -\Lambda_s \frac{\partial f_b}{\partial \Lambda_s} \tag{3.130}$$

one obtains a relation between the solvent fugacity and the density given by

$$\Lambda_s = \rho_{sb} \exp \left(\frac{Q^2 \varepsilon_0 \ell_B a^2 \Lambda^3}{12\varepsilon_w} \right). \tag{3.131}$$

Inserting the expression for $v_d(z, a_z)$ into Eq. (3.124), making use of Eqs. (3.126), (3.127) and the fugacity (3.131), one finally obtains the dipolar term of the grand potential given by

$$\Omega_d = -S\rho_{sb} \int\limits_0^d dz \int\limits_{a_1(z)}^{a_2(z)} \frac{da_z}{2a} e^{-\frac{Q^2}{2}\delta v_d(z, a_z)}, \tag{3.132}$$

where the argument in the exponential function is the dipolar self-energy given by the fairly involved expression

$$\delta v_d(z, a_z) = \frac{\Lambda^3 \ell_B}{6} \left\{ \frac{\varepsilon_0}{\sqrt{\varepsilon_\parallel \varepsilon_\perp}} \left(a_\parallel^2 + \Gamma a_z^2 \right) - \frac{\varepsilon_0}{\varepsilon_w} a^2 \right\}$$

$$+ \int_0^\Lambda \frac{dk\,k}{2\pi} \left\{ \delta v_0(z, z) + \delta v_0(z + a_z, z + a_z) - 2\delta v_0(z, z + a_z) J_0(ka_\parallel) \right\},$$

$$(3.133)$$

where the function $\delta v_0(z, z')$ is given before by Eq. (3.110).

We thus have obtained our desired full expression for $\Omega_v(\varepsilon_\parallel, \varepsilon_\perp)$. The remaining task is the numerical minimization of the grand potential in Eq. (3.104) with respect to the two variational parameters $\varepsilon_\parallel$ and $\varepsilon_\perp$ for which we use a *dichotomy algorithm* with a UV cut-off set to $\Lambda = 200/\ell_B$. The algorithm is described in detail in Ref. [14].

We now go on to deduce the physics that results from this final heavy exercise. The anisotropic dielectric permittivites are shown in Fig. 3.8a and b. Their non-monotonic behaviour can be explained in terms of a competition between the *image-dipole interactions* and their dielectric screening. According to Eqs. (3.76)–(3.77),

Fig. 3.8 a Dielectric permittivity components renormalized by the reservoir permittivity against the bulk solvent concentration. The pore diameter is fixed at $d = 1$ nm, and the liquid contains no ions ($\rho_{ib} = 0.0$ M). **b** Dielectric permittivity components renormalized by the reservoir permittivity (at a fixed value of $\rho_{sb} = 55$ M, and free of ions $\rho_{ib} = 0.0$ M) as a function of pore diameter d. Reproduced from [14], with the permission of AIP Publishing

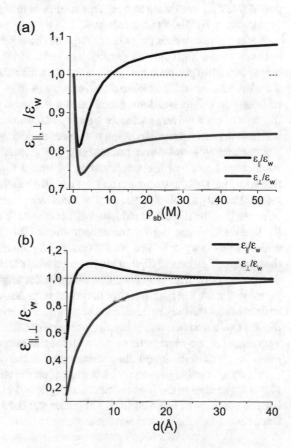

the pore solvent density is fixed by the dipolar self-energy (3.133) obtained from the electrostatic propagator (3.110). Equation (3.110) indicates that the amplitude of the propagator is in turn set by the coefficient

$$\Delta_\Gamma / \sqrt{\varepsilon_\parallel \varepsilon_\perp} \sim \Delta_0 / \varepsilon_w \tag{3.134}$$

where the permittivity ε_w in the denominator determines the intensity of the dielectric screening of image interactions and the amplitude

$$\Delta_0 \equiv \frac{\varepsilon_w - 1}{\varepsilon_w + 1} \tag{3.135}$$

is the *dielectric jump function*. The function Δ_0 / ε_w has its minimum at $\varepsilon_w = 1 + \sqrt{2}$. This corresponds to a bulk density $\rho_{sb}^* \simeq 1$ M. In the density regime $0 \le \rho_{sb} \le \rho_{sb}^*$, the coefficient Δ_0 / ε_w is amplified with an increase of the solvent density ρ_{sb}, because the function Δ_0 rises faster than the permittivity ε_w. Thus, image forces dominate in this regime the dielectric screening. With the increase of the density ρ_{sb} above the characteristic value ρ_{sb}^*, the dielectric jump function Δ_0 saturates to one but the permittivity ε_w continues to rise. This results in turn in a net reduction of the image dipole forces by dielectric screening.

We can also consider the effect of the pore diameter d on the behaviour of confined solvents at the high solvent density $\rho_{sb} = 55$ M. Figure 3.8b displays the anisotropic dielectric permittivities against the pore diameter varied between pore sizes of $d = 1$ Å and 40 Å. The dielectric anisotropy in the slit pore is a direct consequence of image dipole interactions. Since the negative and positive elementary charges at the ends of each solvent molecule are subject to the same image forces, image dipole interactions favor the polarization of solvent molecules along the membrane surface. This results in a transverse permittivity that exceeds the longitudinal component. As a consequence of the amplification of image dipole interactions with the confinement, the solvent exclusion and the resulting dielectric reduction effects become relevant mostly at sub-nanometer pore sizes. Moreover, as the pore size is decreased, one sees that the dielectric reduction effect occurs in a more pronounced fashion for the longitudinal than the transverse component. The former decreases monotonically until it drops to $\varepsilon_\perp \simeq 0.2 \, \varepsilon_w$ for the (extreme) pore diameter $d = 1$ Å. This is due to the combined effects of dipolar alignment and exclusion that are both amplified with decreasing pore size. However, since the dipolar alignment increases the transverse permittivity, we see that the latter first rises with decreasing pore size and then starts to drop below the bulk permittivity below a characteristic pore size $d \simeq 5$ Å when the dipolar exclusion dominates the alignment effect. The dielectric anisotropy becomes very large for sub-nanometer pores, with the transverse permittivity being about four times larger than the longitudinal component at the pore size $d = 1$ Å.

We conclude this section with a comment on the *van der Waals-interactions*. They play a major role in the interactions between low dielectric bodies, e.g., in determining the stability of large colloids [21]. The standard formulation of van der Waals forces considers the solvent surrounding the colloids as a dielectric continuum liquid whose dielectric permittivity is unaffected by the presence of macromolecules. However, the previous results show that this assumption is incorrect for sub-nanometer intermolecular distances that become comparable to the solvent molecular size. Motivated by this observation, we evaluate the impact of the dielectric anisotropy on the interaction between the neutral walls. We recall from Chap. 1 that the interaction force between the interfaces is given by the derivative of the total free energy (3.104) with respect to the interplate distance, that is

$$\Pi(d) = -\frac{1}{S}\frac{\partial \Omega_v}{\partial d}. \tag{3.136}$$

Since we would like to make a qualitative comparison with the standard van der Waals-forces, we will consider exclusively the van der Waals part of the free energy corresponding to the last term of Eq. (3.121). Taking the limit of infinite cut-off $\Lambda \to \infty$, carrying out the integral in Fourier space and differentiating the van der Waals free energy with respect to the pore size d, the interaction force follows in the form

$$\Pi_{vdW} = -\frac{A}{8\pi d^3}, \tag{3.137}$$

where we introduced an effective *Hamaker constant*

$$A = \frac{\varepsilon_\perp}{\varepsilon_\parallel} Li_3 \left(\Delta_\Gamma^2 \right). \tag{3.138}$$

In Eq. (3.138), $Li_3(x)$ is the *polylogarithm function of third order* [22]. The standard van der Waals-interaction of the dielectric continuum formulation is likewise given by Eq. (3.137) but with a different Hamaker constant $A \to A_0 = Li_3(\Delta_0^2)$ which is recovered in the limit $\varepsilon_{\parallel,\perp} \to \varepsilon_w$, with the coefficient Δ_0 given by Eq. (3.135) [21].

A comparison of the interaction force in Eq. (3.137) with the usual van der Waals pressure shows that for interplate separations in the range $d \lesssim 1$ nm, the present theory predicts a significantly lower pressure than the van der Waals theory. To understand this point, we note that the dielectric anisotropy and dipolar exclusion effects result in the inequalities $\varepsilon_\perp/\varepsilon_\parallel < 1$ and $\Delta_\Gamma < \Delta_0$, respectively. According to Eq. (3.138), this reduces the effective Hamaker constant below the standard value, that is one has $A < A_0$. We conclude that for intercolloidal distances in the sub-nanometer range, the dielectric continuum formulation of macromolecular interactions that neglects solvent-membrane interactions overestimates the van der Waals attraction between dielectric bodies.

3.10 Differential Capacitance

As a final result for the nonlocal variational model we discuss the effect of non-local correlations on the *differential capacitance* of an electrolyte in contact with a charged surface located at $z = 0$. The knowledge of the dielectric response of water in the vicinity of low polarity substrates is important for the optimization of new generation energy devices such as *supercapacitors*, which are usually fabricated from carbon-based materials with low static permittivities on the order $\varepsilon_m \simeq 2 - 5$, hence they are ideal realizations of the limit $\varepsilon_m \ll \varepsilon_w$. The standard theory that allows to predict the charge storage ability of these systems is the *Gouy-Chapman (GC) model*, which is based on the Poisson-Boltzmann formalism [23]. Within the framework of this mean-field theory, the differential capacitance is defined as

$$C = \frac{q_i e^2}{k_B T} \left| \frac{\partial \sigma_s}{\partial \phi_0(0)} \right| \tag{3.139}$$

and is given by the simple formula

$$C = \varepsilon_w \kappa_{DH}, \tag{3.140}$$

where the Debye-Hückel parameter is defined in Eq. (3.91); see also Chap. 1.

It is well-known that the Guoy-Chapman capacitance overestimates by several factors the experimental capacitance data of materials at the point-of-zero-charge, see, e.g., Ref. [24]. To evaluate the importance of non-locality on the differential capacitance, we first compare the potential profiles obtained from the mean-field Poisson-Boltzmann and NLPB equations, see Fig. 3.9a. As a result of the thin interfacial dielectric reduction layer the NLPB potential at the surface is only slightly below the Poisson-Boltzmann result. In Fig. 3.9b, one notices that this lowers the Poisson-Boltzmann capacitance, an effect that becomes more pronounced with increasing salt concentration but remains still not sufficient to explain the experimental observations.

By contrast, the electrostatic potential obtained from the solution of Eq. (3.80) is lower than the NLPB result by several factors, and remains very close to the result of the local EDPB model. This stems again from the dielectric screening deficiency associated with the large dielectric exclusion layer in Fig. 3.9a. Finally, in Fig. 3.9b, the same interfacial dielectric reduction amplified by image dipole interactions is seen to drop the differential capacitance of the mean-field NLPB formalism to the order of magnitude of the experimental data that lie in the range of values yielded by the variations models and hence substantially improve over the Gouy-Chapman capacitance; for more details, we refer to [14]. The comparison of the non-local and EDPB results indicates that non-locality only brings a minor contribution to the capacitance. This point confirms that the interfacial solvent depletion induced by image interactions plays the leading role in the low amplitudes of the differential capacitances at the point-of-zero-charge.

Figure 3.9b also contains experimental data, taken from Ref. [24]. As is clearly visible in the Figure, the data follow clearly the trend of the variational models, which gives strong support for the depletion effect captured by these models.

In order to end the discussion of the differential capacitance, we now go back to a simple model, exploiting these insights before concluding on the insights gained in this chapter.

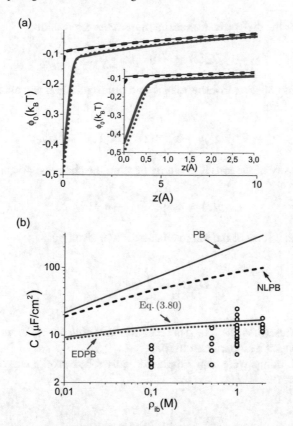

Fig. 3.9 **a** Electrostatic potential profile $\phi_0(k_B T)$ and **b** differential capacitance $C(\mu\text{F/cm}^2)$ against the bulk salt concentration for the solvent with reservoir density $\rho_{sb} = 50.8$ M in contact with a planar interface, with the membrane permittivity $\varepsilon_m = 1$. In **a**, the interfacial charge is $\sigma_s = 0.01$ e nm^{-2} and the salt density $\rho_{ib} = 0.1$ M. The capacitances in **b** correspond to the point-of-zero-charge reached in the limit $\sigma_s \to 0.0$. The comparison between the results from the four different models is discussed in the main text, as well as the discussion of the experimental data, plotted in **b**. Reproduced from [14], with the permission of AIP Publishing

3.11 Going Simple Again: A Phenomenological Model for the Differential Capacitance

In order to assess the role of the *dielectric void* occurring as a consequence of the image forces near the interface we consider a minimal, analytically solvable model by Netz and collaborators in [25]. Considering here only the variation of the permittivity in the perpendicular direction, the Maxwell equation corresponds to the Poisson-Boltzmann equation

$$\nabla \varepsilon_0 \varepsilon_\perp(z) \nabla \psi(z) = -\varrho(z),\tag{3.141}$$

i.e., we have for the dielectric displacement field the expression

$$D_\perp(z) = -\varepsilon_0 \varepsilon_\perp(z) \nabla \psi(z). \tag{3.142}$$

Rearranging this equation in terms of $\nabla \psi$ and differentiating, one obtains a Poisson equation of the form

$$\varepsilon_0 \nabla^2 \psi(z) = -\varepsilon_\perp^{-1} \varrho(z) - D_\perp(z) \nabla \varepsilon_\perp^{-1}(z). \tag{3.143}$$

The charge density of the ions is given by (setting the chemical potential to zero)

$$\varrho(z) = ec_0 [e^{-\frac{e\psi(z)}{k_B T}} - e^{\frac{e\psi(z)}{k_B T}}], \tag{3.144}$$

and the integrated form of the Maxwell equation is given by

$$D_\perp(z) = \int_{-\infty}^{z} dz' \varrho(z'). \tag{3.145}$$

The system of Eqs. (3.143)–(3.145) is set of *nonlinear integro-differential equations*, requiring in general a numerical solution.

In Ref. [25], things were kept simple and only a box profile was considered for $\varepsilon_\perp(z)$ of the form

$$\varepsilon_\perp(z) = \begin{cases} 1, & \text{for } z \leq z^* \\ \varepsilon, & \text{for } z > z^*. \end{cases} \tag{3.146}$$

From this construction we can immediately obtain

$$\nabla \varepsilon_\perp^{-1} = -\left(1 - \frac{1}{\varepsilon}\right) \delta(z - z^*), \tag{3.147}$$

and from Eq. (1.115) we find after integration the boundary condition at $z^* = 1$:

$$\varepsilon_0 [\nabla \psi(z^* + 0) - \nabla \psi(z^* - 0)] = \left(1 - \frac{1}{\varepsilon}\right) D_\perp(z^*) \tag{3.148}$$

with the first term on the right-hand side of Eq. (3.143) vanishing after integration since $\varepsilon = 1$ at $z^* = 1$. The box model therefore simply results in the discussion of a Poisson-Boltzmann problem at the point $z = z^*$ with different dielectric constants in the subspaces $[0, z^*]$ and (z^*, ∞). The surface charge located at $z = 0$ leads to the boundary condition

$$\varepsilon_0 \nabla \psi(z = 0) = -\sigma_0. \tag{3.149}$$

In the next step we linearize the Boltzmann equation part

$$\varrho(z) \approx -\frac{2e^2 c_0 \psi(z)}{k_B T} \qquad (3.150)$$

so that we obtain

$$\nabla^2 \psi(z) = \begin{cases} \varepsilon \kappa^2 \psi(z) & \text{for } z \leq z^* \\ \kappa^2 \psi(z) & \text{for } z > z^*. \end{cases} \qquad (3.151)$$

The Debye-Hückel screening length κ^{-1} is defined by

$$\kappa^2 = \frac{2e^2 c_0}{\varepsilon_0 \varepsilon k_B T}, \qquad (3.152)$$

and the general solution of the linearized PB equation is given by

$$\psi(z) = \begin{cases} A e^{-\sqrt{\varepsilon}\kappa z} + B e^{\sqrt{\varepsilon}\kappa z} & \text{for } z \leq z^* \\ F e^{-\kappa z} & \text{for } z > z^*. \end{cases} \qquad (3.153)$$

Using the boundary and continuity conditions from above one obtains for the amplitudes A, B and F the expressions

$$A = \frac{\sigma_0}{\varepsilon_0 \sqrt{\varepsilon}\kappa} \left(e^{-2\sqrt{\varepsilon}\kappa z^*} \left(\frac{\sqrt{\varepsilon}-1}{\sqrt{\varepsilon}+1} \right) + 1 \right)^{-1} \qquad (3.154)$$

$$B = -\frac{\sigma_0}{\varepsilon_0 \sqrt{\varepsilon}\kappa} \left(e^{2\sqrt{\varepsilon}\kappa z^*} \left(\frac{\sqrt{\varepsilon}+1}{\sqrt{\varepsilon}-1} \right) + 1 \right)^{-1} \qquad (3.155)$$

$$F = A e^{(1-\sqrt{\varepsilon})\kappa z^*} + B e^{(1+\sqrt{\varepsilon})\kappa z^*}. \qquad (3.156)$$

The differential surface capacitance defined by

$$C = \left(\frac{d\psi(z=0)}{d\sigma_0} \right)^{-1}, \qquad (3.157)$$

i.e. the change of the electrostatic potential at the surface with respect to surface charge σ_0 can be obtained from the solution for $\psi(z)$

$$C = \varepsilon_0 \sqrt{\varepsilon}\kappa \frac{1 + e^{-2\sqrt{\varepsilon}\kappa z^*}(\sqrt{\varepsilon}-1)/(\sqrt{\varepsilon}+1)}{1 - e^{-2\sqrt{\varepsilon}\kappa z^*}(\sqrt{\varepsilon}-1)/(\sqrt{\varepsilon}+1)}. \qquad (3.158)$$

It is instructive to consider the asymptotic limits of this involved expression. In the limit $z^*\kappa \to 0$ which corresponds to a uniform bulk dielectric constant, one finds

$$C \approx \varepsilon_0 \varepsilon \kappa \tag{3.159}$$

i.e. the standard capacitance for a system with dielectric constant ε and a capacitor thickness κ^{-1}. In the opposite limit $z^*\kappa \to \infty$, which corresponds to the case when the double-layer width κ^{-1} is much smaller than the interfacial layer width z^*, one finds

$$C \approx \varepsilon_0 \sqrt{\varepsilon} \kappa \tag{3.160}$$

which corresponds to a capacitor of plate separation $1/(\kappa\sqrt{\varepsilon})$. Further, to linear order in κz^*, one has

$$C \approx \varepsilon_0 \varepsilon \kappa \left(1 - \kappa z^*(\varepsilon - 1)\right), \tag{3.161}$$

which can be rewritten as

$$C \approx \left(\frac{\kappa^{-1} - z^*}{\varepsilon_0 \varepsilon} + \frac{z^*}{\varepsilon_0}\right)^{-1}. \tag{3.162}$$

This expression can be read as the capacitance of two capacitors in series: alongside with the interfacial capacitor of thickness z^* with the permittivity ε_0 of vacuum one has the outer double layer of thickness $\kappa^{-1} - z^*$ with bulk permittivity $\varepsilon_0 \varepsilon$. The comparison of the exact expression, Eq. (3.158), with the asymptotic limits, Eqs. (3.159), (3.160), shows that the values lie in the same range at high salt concentrations, see 3.9b and Fig. 1 in Ref. [25]. The main message of this model is thus that this simple theory explains—because of the $\sqrt{\varepsilon}$ behaviour—why the experimental data typically lie one or two orders of magnitude below the standard capacitance.

While this model is in accord with experiment, we however do not really go deep in an understanding of the physics underlying the problem. The more detailed and richer fluctuation theory we developed before on the basis of the variational approach to Poisson-Boltzmann theory has provided this deeper insight into the problem—in particular through the comparison of different model versions, as this allows to better pinpoint the exact reason for the results found in experiment.

3.12 Summary

In this chapter we have discussed how to build explicit solvent models for Poisson-Boltzmann theory. We developed essentially two approaches; the first, a phenomenological one, is *nonlocal electrostatics*. Under this notion is quite generally understood that the dielectric behaviour of a solvent, and in particular of water, cannot be captured by its static dielectric constant, as is the case in the standard Poisson-Boltzmann equation. This is illustrated in Fig. 3.1, which displays the full dielectric function of water as a function of wave-vector, $\varepsilon(k)$. In order to remedy this feature of the

Poisson-Boltzmann equation, we briefly discussed this phenomenological approach to this problem that has been pioneered by A. A. Kornyshev and his collaborators.

Subsequently, we presented a statistical physics approach to the problem. Starting from a model of molecular dipoles of finite size that are caricature-versions of real water molecules, we derived the nonlocal mean-field Poisson-Boltzmann equation and solved it in a simple planar geometry. This calculation allowed us to explicitly compute the space and wave-vector dependence of the dielectric function $\varepsilon(k)$.

Following this first step we have formulated a *fluctuation-corrected Poisson-Boltzmann equation* by employing a variational method. The Hamiltonian of the charged system is averaged with a Boltzmann distribution with a general quadratic form. Illustrated first for a local Hamiltonian in the previous chapter, the procedure is then performed in detail on the dipolar Poisson-Boltzmann Hamiltonian. With the equations now at hand, we could proceed to derive predictions from our theory. We derived the non-local self-consistent Eqs. (3.65) and (3.68) for the charged liquid in an arbitrary geometry. These equations which take into account the extended charge structure of the solvent molecules generalize the local variational equations of Netz and Orland from Ref. [17]. The non-local equations were first solved to determine the dielectric permittivity, allowing for direct comparison with mean-field results and local variational models. Correlation effects on the interfacial dielectric response of the solvent at simple charged planes. We showed that the major effect of correlations is the decrease of the MF surface polarization charge densities, resulting in a reduced dielectric permittivity in the vicinity of the charged interface.

For a dilute solvent confined in a nanoslit we then showed that the explicit consideration of the solvent in the variational approach already results in ionic image-charge interactions responsible for ion rejection from membrane nanopores that are seen in experiment [26]. Furthermore, the difference between the solvent-explicit and local image-charge potentials was shown to result from the additional ionic Born energy between the pore and the ion reservoir. We also found that the latter becomes important exclusively for slits of subnanometer size. This observation allows to fix the confinement scale where the dielectric continuum electrostatics is expected to be valid.

We then considered the solvent model at physiological concentrations within a restricted variational approach based on a dielectrically anisotropic kernel. It was found that the interaction of solvent molecules with their electrostatic images results in two effects. First of all, the dipolar alignment along the membrane surface leads to an anisotropic dielectric response of the fluid characterized by a transverse permittivity exceeding the longitudinal permittivity, i.e. $\varepsilon_\parallel > \varepsilon_\perp$. The dielectric reduction and anisotropy become relevant for pores of subnanometer size, and both effects were shown to reduce the Hamaker constant of the standard van der Waals theory. This result indicates that the dielectric continuum formulation of macromolecular interactions overestimates the van der Waals attraction between dielectric bodies at small separations.

For the dielectric capacitance of an interface, the interfacial dielectric void induced by dipole-image interactions was shown to largely dominate the mean-field dielectric reduction associated with non-locality. We also compared with experimental capacitance data the prediction of the present non-local theory with and without correlations. We found that the mean-field theory including only solvent structure is not sufficient to explain the low values of the experimental data, and the consideration of correlations responsible for the strong interfacial solvent depletion is necessary to have a quantitative agreement with experiments. Finally, we presented a simple mean-field model rationalizing this effect.

3.13 Further Reading

As said in the Summary, the phenomenological approach to nonlocal electrostatics has been pioneered by A. A. Kornsyhev and his collaborators. In a series of early papers, the nonlocal Poisson-Boltzmann equation was solved for the simple geometries we discussed in Chap. 1, for different choices of phenomenological dielectric functions [4–6].

Rather than handling an unwieldy nonlocal Poisson-Boltzmann equation, one can equivalently describe nonlocality in terms of polarization functionals. This approach dates again back to Kornyshev [2], but has enjoyed some ongoing interest in the works by Maggs et al. [27–33].

Generalizations of the Poisson-Boltzmann equation to explicit solvents of dipolar type have been particularly developed by H. Orland, D. Andelman and collaborators. In their works, various versions of dipolar models have been derived and studied, but generally by assuming that the solvent dipoles are point dipoles. As we saw, the Poisson-Boltzmann equation is then turned into an equation with an additional nonlinearity, it however remains a local expression. Exemplary studies are [9, 34–38]. A nice overview has been written by Frydel [39].

Most recently, the nonlocal approach has e.g. found the interest of R. Netz and colloborators, who have built a Poisson-Boltzmann modeling approach on the basis of very detailed molecular dynamics (MD) simulations. Our discussion in the last section of this chapter stems from this line of research. Some relevant papers are [24, 40–43].

A recent literature review on the differential capacitance in the context of ionic liquids is [44].

References

1. Bopp, P.A., Kornyshev, A.A., Sutmann, G.: Static nonlocal dielectric function of liquid water. Phys. Rev. Lett. **76**, 1280–1283 (1996)
2. Kornyshev, A.A., Leikin, S., Sutmann, G.: "Overscreening" in a polar liquid as a result of coupling between polarization and density fluctuations. Electrochim. Acta **42**, 849–865 (1997)
3. Bopp, P.A., Kornyshev, A.A., Sutmann, G.: Frequency and wave-vector dependent dielectric function of water: collective modes and relaxation spectra. J. Chem. Phys. **109**, 1939–1958 (1998)

4. Kornyshev, A.A., Rubinshtein, A.I., Vorotyntsev, M.A.: Model nonlocal electrostatics: I. J. Phys. C: Solid State Phys. **11**, 3307–3322 (1978)
5. Vorotyntsev, M.A.: Model nonlocal electrostatics: II. Spherical interface J. Phys. C: Solid State Phys. **11**, 3323-3331 (1978)
6. Kornyshev, A.A., Vorotyntsev, M.A.: Model nonlocal electrostatics: III. Cylindrical interface J. Phys. C : Solid State Phys. **12**, 4939-4946 (1979)
7. Hildebrandt, A., Blossey, R., Rjasanow, S., Kohlbacher, O., Lenhof, H.-P.: Novel formulation of nonlocal electrostatics. Phys. Rev. Lett. **93**, 108104 (2004)
8. Hildebrandt, A., Blossey, R., Rjasanow, S., Kohlbacher, O., Lenhof, H.-P.: Electrostatic potentials of proteins in water: a structured continuum approach. Bioinformatics **23**, e99–e103 (2006)
9. Abrashkin, A., Andelman, D., Orland, H.: Dipolar Poisson-Boltzmann equation: ions and dipoles lose to charge interfaces. Phys. Rev. Lett. **99**, 077801 (2007)
10. Buyukdagli, S., Blossey, R.: Nonlocal and nonlinear electrostatics of a dipolar Coulomb fluid. J. Phys.: Condens. Matter **26**, 285101 (2014). https://doi.org/10.1088/0953-8984/26/28/285101
11. Mallik, B., Masunov, A., Lazaridis, T.: Distance and exposure dependent effective dielectric function. J. Comput. Chem. **11**, 1090–1099 (2002)
12. Buyukdagli, S., Manghi, M., Palmeri, J.: Variational approach for electrolyte solutions: from dielectric interfaces to charged nanopores. Phys. Rev. E **81**, 041601 (2010)
13. Buyukdagli, S., Achim, C.V., Ala-Nissila, T.: Electrostatic correlations in inhomogeneous charged fluids beyond loop expansion. J. Chem. Phys. **137**, 104902 (2012)
14. Buyukdagli, S., Blossey, R.: Dipolar correlations in structured solvents under nanoconfinement. J. Chem. Phys. **140**, 234903 (2014). https://doi.org/10.1063/1.4881604
15. Coalson, R.D., Duncan, A.: Statistical mechanics of a multipolar gas: a lattice field theory approach. J. Phys. Chem. **100**, 2612–2620 (1996)
16. Buyukdagli, S., Ala-Nissila, T.: Dipolar depletion effect on the differential capacitance of carbon-based materials. EPL **98**, 60003 (2012)
17. Netz, R.R., Orland, H.: Variational charge renormalization in charged systems. Eur. Phys. J. E **11**, 301–311 (2003)
18. Buyukdagli, S., Ala-Nissila, T.: Microscopic formulation of nonlocal electrostatics in polar liquids embedding polarizable ions. Phys. Rev. E **87**, 063201 (2013)
19. Netz, R.R.: Static van der Waals interactions in electrolytes. Eur. Phys. J. E **5**, 189–205 (2001)
20. Ballenegger, V., Hansen, J.P.: Dielectric permittivity profiles of confined polar fluids. J. Chem. Phys. **122**, 114711 (2005)
21. Israelachvili, J.: Intermolecular and Surface Forces. Academic Press (1992)
22. Abramowitz, M., Stegun, I.A.: Handbook of Mathematical Functions. Dover Publications, New York (1972)
23. Butt, H.-J., Kappl, M.: Surface and Interfacial Forces. Wiley-VCH Verlag GmbH & Co. KGaA, Weinheim, Germany (2010)
24. Bonthuis, D.J., Gekle, S., Netz, R.R.: Dielectric profile of interfacial water and its effect on double-layer capacitance. Phys. Rev. Lett. **107**, 166102 (2011)
25. Bonthuis, D.J., Uematsu, Y., Netz, R.R.: Interfacial layer effects on surface capacitances and electro-osmosis in electrolytes. Phil. Trans. R. Soc. A **374**, 20150033 (2015)
26. Tu, C.-H., Fang, Y.-Y., Zhu, J., van der Bruggen, B., Wang, X.-L.: Free energies of the ion equilibrium partition of KCl into nanofiltration membranes based on transmembrane electrical potential and rejection. Langmuir **27**, 10274–10281 (2011)
27. Maggs, A.C., Everaers, R.: Simulating nanoscale dielectric response. Phys. Rev. Lett. **96**, 230603 (2006)
28. Paillusson, F., Blossey, R.: Slits, plates, and Poisson-Boltzmann theory in a local formulation of nonlocal electrostatics. Phys. Rev. E **82**, 052501 (2010)
29. Berthoumieux, H., Maggs, A.C.: Fluctuation-induced forces governed by the dielectric properties of water–a contribution to the hydrophobic interaction. J. Chem. Phys. **143**, 104501 (2015)
30. Berthoumieux, H.: Gaussian field model for polar fluids as a function of density and polarization: Toward a model for water. J. Chem. Phys. **148**, 104504 (2018)

31. Berthoumieux, H., Paillusson, F.: Dielectric response in the vicinity of an ion: a nonlocal and nonlinear model of the dielectric properties of water. J. Chem. Phys. **150**, 094507 (2019)
32. Monet, G., Bresme, F., Kornyshev, A.A., Berthoumieux, H.: Nonlocal dielectric response of water in nanoconfinement. Phys. Rev. Lett. **126**, 216001 (2021)
33. Berthoumieux, H., Monet, G., Blossey, R.: Dipolar Poisson models in a dual view. J. Chem. Phys. **155**, 024112 (2021)
34. Koehl, P., Orland, H., Delarue, M.: Computing ion solvation free energies using the dipolar Poisson model. J. Phys. Chem. B **113**, 5694–5697 (2009)
35. Koehl, P., Orland, H., Delarue, M.: Beyond the Poisson-Boltzmann model: modeling biomolecules-water and water-water interactions. Phys. Rev. Lett. **102**, 087801 (2009)
36. Lévy, A., Andelman, D., Orland, H.: Dielectric constant of ionic solutions: a field-theory approach. Phys. Rev. Lett. **108**, 227801 (2012)
37. Koehl, P., Poitevin, F., Orland, H., Delarue, M.: Modified Poisson-Boltzmann equation for characterizing biomolecular solvation. J. Theor. Comp. Chem. **13**, 1440001 (2014)
38. Poitevin, F., Delarue, M., Orland, H.: Beyond Poisson-Boltzmann: numerical sampling of charge density fluctuations. J. Phys. Chem. B **120**, 6270–6277 (2016)
39. Frydel, D.: Mean-field electrostatics beyond the point-charge description. Adv. in Chem. Phys. **160**, 209–260 (2016)
40. Bonthuis, D.J., Gekle, S., Netz, R.R.: Profile of the static permittivity tensor of water at interfaces: consequences for capacitance, hydration interaction and ion adsorption. Langmuir **28**, 7679–7694 (2012)
41. Bonthuis, D.J., Gekle, S., Netz, R.R.: Unraveling the combined effects of dielectric and viscosity profiles on surface capacitance, electro-osmotic mobility, and electric surface conductivity. Langmuir **28**, 16049–16059 (2012)
42. Bonthuis, D.J., Netz, R.R.: Beyond the continuum: how molecular solvent structure affects electrostatics and hydrodynamics at solid-electrolyte interfaces. J. Phys. Chem. B. **117**, 11397–11413 (2013)
43. Bonthuis, D.J., Netz, R.R.: Dielectric profiles and ion-specific effects at aqueous interfaces. In: Dean et al. (eds.) Electrostatics of Soft and Disordered Matter, pp. 129–142. Pan Stanford Publishing, Singapore (2014)
44. Budkov, Y.A., Kolesnikov, A.L.: Electric double layer theory for room temperature ionic liquids on charged electrodes: milestones and prospects. Curr. Op. Electrochem. **33**, 100931 (2022)

Chapter 4
Conclusions

In this book we have discussed the Maxwell equation

$$\nabla \cdot \mathbf{D} = \varrho(\mathbf{r}) \tag{4.1}$$

in which we modified the left-hand and the right-hand side of the equation in several ways to treat a key problem of soft matter electrostatics: the properties of confined electrolytes.

The simplest rewriting of the Maxwell equation in the context of confined electrolytes is the 'standard' Poisson-Boltzmann equation, in which the left-hand side of Eq. (4.1) is turned into the Laplace operator acting on the electrostatic potential, while the right-hand side consists of Boltzmann factors. We have solved this equation for elementary geometries (planar, cylindrical, spherical).

Subsequently, we have seen that the Poisson-Boltzmann equation can be generalized for fluid systems described by a general class of equations of state; in particular, under the assumption of relevant spatial variations of the underlying fluid, it can even turn into a differential equation of higher order.

In the next step, we have derived the Poisson-Boltzmann equation from a statistical mechanics approach by which we calculated the grand partition function for a system of Coulomb charges in a saddle-point approximation. In this approach, the saddle-point (mean-field) equation yields the Poisson-Boltzmann equation while the quadratic fluctuations around the saddle-point determine the so-called one loop correction to the grand potential. We looked at this for the determination of the effect of dissolved ions on the surface tension of an air-water interface.

Following these key insight, we have moved on to relax a key assumption of the Poisson-Boltzmann equation, namely that the dielectric properties of the system are dominated by the static dielectric constant. We have first studied the phenomenological approach to this problem, in which the Poisson-Boltzmann equation turns into an

© The Author(s), under exclusive license to Springer Nature Switzerland AG 2023   97
R. Blossey, *The Poisson-Boltzmann Equation*,
SpringerBriefs in Physics, https://doi.org/10.1007/978-3-031-24782-8_4

integro-differential equation. Furthermore, we have studied the statistical mechanics of a solvent made up of finite size dipoles. For this system we have discussed the mean-field Poisson-Boltzmann equation as well as developed a second approach which allows to treat fluctuations beyond mean-field behaviour that is based on a variational approach which we had already introduced in Chap. 2. The physics that can be derived from such an approach for an electrolyte filling a nanoslit has been discussed in quite some detail.

The three Chapters of the book have led through a number of key approaches in the development of the Poisson-Boltzmann equation. In the Introduction the question was posed: is the Poisson-Boltzmann approach a *theory* or is it a modeling approach?

I hope that I have given an answer to this question. In going from simpler to more complex, refined models like the fluctuation-corrected and even nonlocal Poisson-Boltzmann models we have, not least by comparing different levels of theoretical treatments, we have been able to pinpoint a number of physical features. Comparing them to experiment then explains the relevance of different effects. Finally, we have shown that these insights can be used to derive very simplified Poisson-Boltzmann models that allow for a quick comparison to experiment.

These results can be taken as good evidence for the fact that Poisson-Boltzmann theory is *both* a proper theory and a very pragmatic modeling approach. Our more detailed investigation in Chap. 3 that contains a lot of physical detail could be later in part rationalized in terms of a phenomenological mean-field theory produced a nice agreement with experimental results. Taken as such, the box model seems to be pulled out of the hat, but in fact it has a physical underpinning which a more detailed Poisson-Boltzmann result can provide. Poisson-Boltzmann theory is thus *both* a proper theory—and a workhorse approach for modeling and comparison to experiment.

Index

Printed in the United States
by Baker & Taylor Publisher Services